THE PENOBSCOT EXPEDITION

A Colonial America Naval Disaster

The story of the 1779
American Revolutionary War Battle
to prevent the creation of the

British Crown colony of

New Ireland

in Maine

*We may have lost the battle but
we won the war*

ISBN 978-1-935616-17-7

For additional copies or more information, please contact:

Salty Pilgrim Press
17 Causeway Street
Millis, MA 02054 USA
1 508 794-1200
captain@saltypilgrim.com

First Edition

Printed in the USA

Forward

Was I ever surprised, having been coming to the Penobscot River region since the 1950s plus owning property on Verona Island at the mouth of the Penobscot River, to learn of American patriot's terrific Naval defeat in 1779 at the hands of the British during the revolutionary War.

I would have had a front row seat from which to watch the battle unfold had I been sitting on my Verona beach on August 14 in 1779!

Ship's timber emerging from Penobscot River mud?

I was even more surprised to learn that my grandparent's farm would be in "New Ireland" – not in Maine or New England, if the American Patriots had not won both the Revolutionary War and the War of 1812.

My grandparent's property fronts on the Penobscot River and a few years ago, thanks to Google Earth, I was able to detect the outline of a sunken ship in the mud at the river's edge.   Could it be one of the 44 ships of the Expedition?

Researching and assembling the information to be presented here was educational, interesting and very enjoyable. I sincerely hope you will find it so as well.

Photo by Deborah Copp Bouchard

TABLE OF CONTENTS

Introduction	1
Background	2
Plans to establish New Ireland	4
British arrive at Castine	8
Fort at Castine a military benefit	9
Personal and political gain	10
Push to establish New Ireland	11
The British command	13
General Sir Henry Clinton	14
General Francis McLean	18
Captain Henry Mowat	23
Loyalist Dr. John Cialef	25
John Nutting	28
The troops	29
Factors that influenced New England's abilty to wage a military campaign	31
The evolution of the Patriot's militia system	32
Military system evolves	35
Town militias organize into 3 statewide units	37
King Philip's War and expeditionary experimentation	38
Bagaduce Peninsula raid starts a French/Indian war	39
Massachusetts gains military confidence	41
Colonist supply the men, British leadership	44
The Penobscot Expedition begins	45
British land at Castine	46
Massachusetts reacts	48
The American command is assembled	51
Commodore Dudley Saltonstall	52
Brigadier-General Peleg Wadsworth	57
Brigadier General Solomon Lovell	61
Lieutenant Colonel Paul Revere	64

Dr. Eliphalet Downer	72
Capt. Jeremiah Hill	73
Joseph McLellan	77
Remainder of the command	78
Massachusetts General Court notified of British occupation	79
Ground forces formed	82
Assembling expeditions	84
Preparing the Provincial Army	86
Captain W. T. Welch Led the Marines	90
Troop recruitment	91
Meanwhile in Castine	95
Patriots planning the attack	98
The American landing	107
A siege instead of victory	114
British reinforcements on the way	117
General Lovell refused to make contingency plans	120
No way the fort could be taken by force	122
Lovell's final attack and the destruction of the American Fleet	126
Saltonstall orders his fleet to Run Away!	128
Aftermath	131
Root cause of failure	133
Penobscot expedition warships	134
British proclamations, journals and letters written by those who were there	135
British officer's summary of the Expedition	136
Proclamation	137
A British journal	141
Captain Mowat's account	160
American proclamations, journals and letters written by those who were there	172
William Moody's journal	173
Proclamation	180
Letter	184
Journal chronicling Col. Mitchell's Regiment's involvement	186

Letter of General Wadsworth regarding   197
the expedition
Col. Enoch Freeman's letter to the   202
Council at Boston
Rev. John Murray's letter   204
Col. Henry Jackson's letter upon arriving   205
at Falmouth August, 1779
Falmouth Selectmen's letter to the   207
Council at Boston
Massachusetts' General Court conducts   209
hearings
Results of the General Court's   210
investigation
Penobscot Expedition opening the   217
campaign
Progress of the siege   222
The defeat   238
Investigatory committee appointed   233
Court's findings   238
Two more attempts to establish New   239
Ireland
Another attempt by the British to   240
establish New Ireland
The Aroostook War of 1842   248
Fort Knox built 1844 - 1869   249
Modern day archaeological – The   251
Penobscot Expedition Site
References   254

It may be not improper to mention that the Action at our landing on Bagaduce might have been called brilliant, had the event of the Enterprise been fortunate. But let military Men not talk of glory who lack success.

– General Peleg Wadsworth - 1828
2nd in command of Patriot fleet

INTRODUCTION

New Ireland became a Crown colony of the United Kingdom in what today is northern Maine after British forces captured the area during the American Revolution and again during the War of 1812.

The colony lasted four years during the Revolution, and eight months during the War of 1812. At the end of each war the United Kingdom ceded the land back to the United States under the Treaty of Paris and Treaty of Ghent, respectively.

The American's Penobscot Expedition was a force confident in its abilities based upon over a century of warfare with the French and eastern Indians. The establishment of a British outpost in Maine was not only a military threat, but it threatened Massachusetts' claim to region.

Maine militiaman

Massachusetts, confident of victory, assembled the Penobscot Expedition without Continental Congress involvement. In most cases, it was justly founded in its abilities, but ultimately lacked the required experience, judgment and training required for success.

The story of the expedition illustrates the ability of the colony, but also highlights inherent flaws in the militia system as well. Coupled with weak and often indecisive leaders, the expedition ended in failure.

1

Background

On July 4th 1776, almost three years earlier, the American civil war of independence from England began. The American War for Independence would last for five more years. Yet to come were the Patriot triumphs at Saratoga, the bitter winter at Valley Forge, the intervention of the French, and the final victory at Yorktown in 1781. In 1783, with the signing of the Treaty of Paris with Britain, the United States formally became a free and independent nation.

Patriots vs. Loyalists

By July 4, 1776 the Patriots had gained control of virtually all territory in the 13 colonies, and expelled all royal officials. No one who openly proclaimed their loyalty to the Crown was allowed to remain, so for the moment, Loyalists fled or kept quiet.

During the American Revolution, those who continued to support King George III of Great Britain came to be known as Loyalists. Loyalists are to be contrasted with Patriots, who supported the Revolution. Historians have estimated that during the American Revolution, between 15 and 20 percent of the population of the colonies, or about 500,000 people, were Loyalists.

Colonial Patriot

As the war concluded with Great Britain defeated by the Americans and the French, the most active Loyalists were no longer welcome in the United States, and sought to move elsewhere in the

2

British Empire. The large majority (about 80%–90%) of the Loyalists remained in the United States, however, and enjoyed full citizenship there.

Historians estimate that a total of 60,000 settlers left the new United States. The majority of them, about 33,000, went to Nova Scotia and New Brunswick.

The departing Loyalists were offered free land in

Loyalist (Tory) mocked by children

British North America. Many were prominent colonists whose ancestors had originally settled in the early 17th century, while a portion were recent settlers in the Thirteen Colonies with few economic or social ties. Many had their property confiscated by Patriots.

Loyalists resettled in Nova Scotia (including modern-day New Brunswick). Their arrival marked the arrival of an English-speaking population in the future Canada west and east of the Quebec border.

In Nova Scotia, there were many Yankee settlers originally from New England, and they generally supported the principles of the revolution. They would remain neutral during the war. During this period Britain had built up powerful forces at their naval base in Halifax, Nova Scotia.

Plans to Establish *New Ireland* - 1777

Following the partially successful British raid on Machias in 1777, as well as General John Burgoyne's failed Saratoga campaign, British war planners looked for other ways to gain control over the rebellious New England colonies however, most of their effort was directed at another campaign with the southern colonies.

Britain's Secretary of State for the Colonies Lord George Germain, the 1st Viscount Sackville, and his Under-Secretary, William Knox, as those responsible for the war effort wanted to establish a base on the coast of the District of Maine (which until achieving statehood in 1820 was a part of Massachusetts) that could be used to protect Nova Scotia's shipping and communities from American privateers and raiders.

Lord George Germain

Among the specific reasons for the British to undertake such an occupation were:

- To protect Nova Scotia's shipping and communities from American privateers and raiders.
- To keep open the timber supply of the Maine coast for masts and spars for the Royal Navy;
- To control the Maine coast down to the Penobscot River (which is adjacent to the Bay of Fundy and Nova Scotia) and to protect the large British naval base at Halifax,

- Because Loyalist refugees had proposed the establishment of a new colony or province to be called New Ireland as a precursor to the establishment of Loyalist New Brunswick in 1784.

In order to promote the idea of establishing a British military presence in Maine, in January 1778 Under Secretary Knox induced John Nutting, a Loyalist who had piloted Sir George Collier's expedition against Machias, to write to Lord Germain and later dispatched him to London to do so in person.

In doing so Nutting described the Castine peninsula as having a harbor that *"could hold the entire British Navy"* and was so easily defendable that *"1,000 men and two ships"* could protect it against any Continental force.

Under-Secretary, William Knox

He also proposed that the strategic location of such a post would help carry the war to New England as well as offer protection for Nova Scotia from attack.

Although Admiral Collier asked Nutting a year later what could possibly have induced him to recommend a settlement there (which he then denied), in light of subsequent events Nutting's figures for defense at Castine proved to be amazingly accurate.

On September 2, 1778, Lord Germain drafted orders for Lieutenant General Sir Henry Clinton, commander-in-chief of British forces in North America (the Colonial Office tasked Nutting to carry said orders to New York) to assist with the establishment of "a province between the Penobscot and St. Croix rivers. Post to be taken on Penobscot River."

Lord Germain's orders to General Sir Clinton read in part:

"The distress of the King's loyal American subjects who have been driven from their habitations and deprived of their property by the rebels has been an object of attention with His Majesty and Parliament from the first appearance of the rebellion; and very considerable sums have been expended in furnishing them with a temporary support. But, as their number is daily increasing and is much to be apprehended (if a reconciliation does not soon take place) that scarcely any who retain their principles will be suffered to remain in the revolted provinces, it is judged proper in that event that a permanent provision should be made by which they may be enabled to support themselves and their families without being a continual burden upon the revenue of Great Britain.

Castine Harbor *"could hold the entire British Navy and 1,000 men and two ships could protect it against any Continental force"*

"The tract of country that lies between Penobscot River and the River St. Croix, the boundary of Nova Scotia on that side, offers itself for the reception of those meritorious but distressed people. And it is the King's intention to erect it into a province. As the first step toward making this establishment it is His Majesty's pleasure, if peace has not taken place and the season of the year is not too far advanced before

you receive this, that you do send such a detachment of troops at Nova Scotia, or of the provincials under your immediate command, as you shall judge proper and sufficient to defend themselves against any attempt the rebels in those parts may be able to make during the winter to take post on Penobscot River, taking with them all necessary implements for erecting a fort, together with such ordnance and stores as may be proper for its defense, and a sufficient supply of provisions."

It was Knox's idea to call this province New Ireland. Unfortunately for the British, an American privateer captured Nutting's ship, and he was forced to dump his dispatches, putting an end to execution of the idea in 1778.

American privateer attacks Nutting's ship

It was against this backdrop that the British's plan to command all of the lands north of the Penobscot River and establish a new colony, the Providence of "New Ireland," unfolded.

British Arrive at Castine

McLean's expedition set sail from Halifax on May 30, 1779 and on 16 June 1779, fourteen year-old William Hutchings, a resident of the Bagaduce peninsula, witnessed the arrival of a large British force into Penobscot Bay.

Castine on the Bagaduce Peninsula

Eight hundred British soldiers under the command of Brigadier General Francis McLean arrived with orders to establish a permanent garrison on the Penobscot River. The next day, the British force began fortifying positions on the peninsula. Anchoring the defenses would be Fort George, located on the central plateau that dominated the peninsula.

The British decision to establish the base on the Penobscot was the product of a series of personal, political, and strategic considerations on both sides of the Atlantic. A base on the Penobscot River brought about serious issues for the Maine Province and its parent colony, Massachusetts. The Bay Colony's rapid response to the crisis would ultimately lack both the ability and the will to drive the British out. In the process, the expedition would nearly financially ruin the colony and further divided the citizens within its eastern province.

8

Fort at Castine a Military Benefit

Establishing an outpost along the Maine coast, although difficult to support, would solve some of Britain's military challenges i.e.:

- It would serve to protect Nova Scotia by both interdicting American privateers that preyed upon British shipping in the Bay of Fundy and,

- it would isolate Machais. Long a haven for revolutionary Nova Scotian refugees Machais had already launched two failed expeditions designed to rally local support against the British.

Privateer attacks along the shipping lanes between Halifax and New York were significant enough to cause Britain to adopt a convoy system to protect its supply lines, thereby drawing off limited naval resources. In a sense, an outpost in Maine would serve as a buffer between New England and Nova

PRIVATEER

- Individual who privately owns a ship
- Government gives them permission in wartime to attack enemy merchant ships
- Sell the cargo and share the money
- American privateers were attacking the British ships
- Could control trade routes and support Patriot cause

Scotia. Additionally, the British politicians recognized that possession of the Penobscot might allow Britain to lay claim to the area if the war ended in a rebel victory.

Personal and Political Gains

As convincing as the military argument may be, the true impetus behind the outpost was both personal and political. Bostonians, John Nutting and Dr. John Calef, both Loyalists loyal to the King of England, saw an opportunity for personal gain in establishing an outpost on the Penobscot River.

Despite the phenomenal growth in Maine's population following the French and Indian War, land grants east of the Penobscot were difficult to gain. Maine's forests provided prized tall, straight naval timbers used to build masts. In fact, the crown generally claimed all such timbers in Maine, but increased settlement east of the Penobscot made enforcement more difficult.

Royal Grant by King George III

Dr. Calef had long sought to gain royal grants for the lands along the Penobscot while Nutting, a successful builder, began to speculate in lands along the Penobscot to build a large lumber enterprise. Both stood to gain immensely if Britain established a foothold in the region and they could legitimize their claims.

As a staunch loyalist, Nutting, was forced to evacuate with the British Army from Boston to Halifax in March of 1776 and resumed his business there. He secured a position as a messenger and travelled to England. Once there, he developed a close relationship with William Knox, the secretary to Lord George Germain, the newly appointed Secretary of State for Colonial Affairs.

Push to Establish New Ireland

Under Secretary Knox and Nutting then developed a plan to rally loyalist support in New England into a new colony, *New Ireland,* within the Maine province. When presented the plan, Lord Germain approved it for both its military merits and as a potential solution to resettlement of the colonies' loyalist population.

Using his role, in developing the overall plan, Nutting had significant input into the selection of a site along the Penobscot and not at a more developed coastal community such as Falmouth. With his plan approved, he departed with the necessary orders for execution in September of 1778.

Nutting's Bad Karma Delays Plans

Unfortunately for Nutting, an American Privateer, the *Vengeance,* intercepted his ship. Wounded four times during the brief battle, Nutting managed to throw his dispatches overboard to prevent their capture. While this dashed any hope of establishing a base prior to the end of the campaign season, fresh dispatches from England arrived in

Nutting wounded in battle

January of 1779. In April, General Henry Clinton in New York sent Nutting to Halifax to give his orders to General Francis McLean.

Clinton orders to McLean were to:

Make such a Detachment of the Troops under your Command, as you shall judge proper and Sufficient to defend themselves against any Attempt the Rebels in those parts may be able to make, directing them to take post on Penobscot River, and sending with them all Necessary Implements for erecting a Fort, together with such Ordnance and Stores as may be proper for its defense, and a sufficient supply of Provisions.

General Clinton estimated that it would take approximately 500 men under one of General Francis McLean's regimental commanders to accomplish this task, but left the final number to General McLean.

Concerned about the likelihood of an American counter-attack, General McLean chose to lead the six hundred and forty men of the 74th Foot and 82nd Foot himself. Attached to his regiments were fifty men the Royal Artillery to man the four twelve, two six, and two four pounder cannon.

The British Command

General Sir Henry Clinton

Revolutionary War Commander-in-Chief of the British Forces

General Sir Henry Clinton, KB (Knight Commander Order of the Bath) 16 April 1730 – 23 December 1795

General Sir Henry Clinton – Commander in Chief

The only son of a British Admiral, Sir Henry Clinton was raised in pre-revolutionary America. His father served as Royal Governor of New York from 1743-1753. Clinton rose steadily through the ranks of the British Army, serving in Germany in the 1760s.

He arrived in **Boston in 1775** and commanded troops under Generals **Thomas Gage** and William Howe at Bunker Hill with some distinction. He proposed and led the double envelopment plan that routed the **Continentals** on **Long Island** in

Clinton led the British at Bunker Hill

1776 as General Howe's second in command. Clinton became Commander in Chief of the British Army in America in upon Howe's recall in 1778 and led his forces to victories at **Monmouth** and **Charleston**. His failure to provide timely aid during the doomed **Yorktown campaign** and led to his resignation in 1781.

Clinton's father was a largely undistinguished Royal Navy officer of an aristocratic family who lived a relatively full life to age seventy-five, dying when Clinton was thirty-one. Clinton's mother was the daughter of a British General who struggled with mental illness as a homemaker. She died when Clinton was thirty-seven

Clinton grew up steeped in the nepotistic world of favor seeking from high hereditary officials from the King on down. He never had a permanent home as a child and never held a job outside of the Army until he served in Parliament from 1774 to 1784. His interpersonal and political skills were weak.

Prior to his return to America in 1775, Clinton had seen action as a senior officer in Germany in the 1760s, serving as aide de camp to the Prince of Brunswick from

1760-62. He was known ever after as a member of the "German School" of continentally experienced British officers. Fluent in French, Clinton was the most cerebral of all the British Generals in America. He kept detailed notes of his years of military reading and was well known for his broad understanding of the political, economic, and geographic context of military policy.

Yet, like a number of other senior British officers, he had never once commanded more than a battalion of men in combat. His instincts for strategy and tactics nonetheless were excellent, he was consistently brave, but his gravitas was lacking.

He was known throughout the service as insightful, but diffident, frequently quarrelsome, impulsive and a loner. Few subordinates liked working for him. Still, he held the confidence of King and Cabinet throughout his service in America.

Clinton served longer in America than any other British Commander. From his first days in Boston in 1775 he sensed the war was a profitless enterprise. He favored a peaceful settlement on almost any terms. He saw war in America for Britain as a business, not a cause.

16

Instead of taking and holding large rebel cities like
New York and later Philadelphia, Clinton believed
the decisive point was George Washington himself
and the Continental Army, justifying energetic and
perhaps costly pursuit. General Howe could not be
convinced of this.

By late 1776, despite his stellar performance

Henry Knighted by the King

during the New York campaign, Clinton realized
his relationship with General Howe had broken
down over philosophic and personal differences.
He returned to England determined to resign his
commission. King George offered him a
knighthood if he would stay on. Clinton
reluctantly agreed, but upon his return to New
York in July 1777, little had changed.

In almost every military situation he faced from
Bunker Hill to the Southern campaigns of 1779-
1781, Clinton suggested the right strategic
approach to either his superiors, his direct
subordinates, and/or his naval counterpart and

every time but one (Long Island), they would not follow his advice. Had General Howe simply followed the original plan for the campaign of 1777 and marched up the Hudson to link up with General John Burgoyne (as Clinton strongly favored) many experts believe the war could have been won. Howe instead marched on Philadelphia and ultimately was relieved of his command. Sir Clinton did lead a smaller force up the Hudson in 1777, but could not break through past Albany in time to relieve Burgoyne.

Upon his promotion to Commander in Chief in May 1778, Clinton found himself matched directly with his temperamental opposite in Washington, a man possessed of almost everything he lacked in personal strengths, even if the Continental Army he commanded was weak. When France entered the war on the American side in February 1778, Clinton sensed he would never have the men, the ships, or the will to prevail over Washington by force.

Clinton sensed he would never have the men, the ships, or the will to prevail over Washington by force.

In 1779, in alignment with the King's wishes, Clinton responded to the French challenge by pursuing a **Southern strategy** that sought to employ modest numbers of British troops in the southern colonies where it was hoped loyalist sympathies ran strongest. Clinton personally led 8,700 troops in a successful assault on Charleston, South Carolina, resulting in the capture of the city on May 12, 1780.

His forces took over 3,300 continental and militia troops prisoner (including seven generals), the worst Continental Army defeat of the war. Initial pacification efforts and amnesties seemed to quell rebellion throughout South Carolina, but over-confidence on Clinton's part led to a final

British surrender at Yorktown

proclamation on June 3 requiring those seeking protection as loyal subjects to take up arms in support of Britain.

This inflamed rebels throughout the colony and led quickly to armed insurrection. Clinton left South Carolina in June with upheaval spreading and ordered his second in command, Lord Charles Cornwallis, to stay put in South Carolina and attempt to solve the problem. Cornwallis essentially ignored this directive and began his own overland campaign into the interior of South Carolina and invaded North Carolina in 1781.

Constantly fearful of an attack by Washington on the weakened garrison at New York (having given up a net 10,000 plus men to other theaters) and harassed by superior numbers of French naval forces off the coast, Clinton was reluctant to dispatch additional troops to Cornwallis in the Carolinas and Virginia in late 1780 and 1781.

Relations and communications again frayed between Britain's top commanders following Clinton's hesitation to resign as planned after Charleston, resulting in the catastrophe at Yorktown.

Clinton was relieved in 1782, returned to England and spent the rest of his life defending his actions in America. He died in 1795.

General Clinton's signature

The Commander at Castine

GENERAL FRANCIS McLEAN

Brigadier General Francis McLean was a sixty-two year old bachelor and veteran of nineteen major battles that spanned Europe, the Caribbean, and Canada. He more recently spent fifteen years in Portugal as a military governor of several fortified cities along the Spanish border. Recalled from Portugal, McLean assumed duties the governor of Nova Scotia and inherited the problems of a growing crowd of Loyalist refugees. He was uniquely qualified to administer the new garrison and potentially a new colony.

General McLean

Brigadier General Francis McLean, also called a Major General, whose name was properly spelled MacLean, was the son of Captain William, who was the grandson of Lachlan, the first of the family of Blaich and second of John Crubach, eighth MacLean of Ardgour. As soon as he was able to carry arms Francis obtained a commission in the same regiment with his father, a regiment of Scottish troops maintained in the Dutch service. He was at the siege of Bergen-op-Zoom in 1747, when the French, after a siege of two months, took the place by storm. "

21

Lieutenants Francis and Allan MacLean (third son of Torloisk) of the Scotch brigade were taken prisoners and carried before General Lowendahl, who thus addressed them: Gentlemen, consider yourself on parole. If all had conducted themselves as you and your brave corps have done, I should not now be master of Bergen-op-Zoom. He was detained prisoner in France for some time; and on his release was promoted to a captaincy and entered the Forty-second Royal Highlanders. At the capture of Gaudaloupe, Francis was severely wounded, but owing to his gallant conduct was promoted to the rank of major and appointed Governor of the island of Marie Galante.

He served in Canada under Wolfe but returned to Great Britain and embarked with the expedition for reducing the island of Belleisle on the coast of France. Here he had his right arm shattered and was taken prisoner. On being exchanged, his bravery was rewarded by promotion to the lieutenant-colonelcy of the 82d. In 1762, he was sent to aid the Portuguese against the combined attack of France and Spain. He was made commander of Almeida, a fortified town on the Spanish frontier, which command he held for several years and was nominated to the government of Estremadura and the City of Lisbon.

On his leaving Portugal in 1778, the king presented him with a handsomely mounted sword, and the queen gave him a valuable diamond ring. On his return to England he was dispatched to America and appointed to the government of Halifax. He repaired with the army in June, 1779, to Penobscot Bay and proceeded to erect defenses.

After the completion of Fort George, McLean and his regiment returned to Halifax where he died, unmarried, May 4, 1781, in his 64th year and was buried two days later.

CAPTAIN HENRY MOWAT

He provided naval support to McLean and commanded the HMS *Albany*. Capt. Mowat was extremely familiar with the intricacies of the Maine coast, serving the last thirteen years either chasing smugglers or privateers. He was infamous amongst the Maine coastal communities for his destruction of Falmouth four years earlier as the captain of the HMS *Canceaux*. Forty-five years old, he had been at sea since he was sixteen. A professional sailor, he took his commitment

Captain Henry Mowat

to the defense of McLean's outpost seriously.

Originally his fourteen-gun *Albany* was the only sloop-of-war to remain behind to support McLean's outpost. The other two ships the *North* and the *Nautilus* had orders elsewhere. In a letter to General Clinton, he described his ship the *Albany* as *"the worst calculated of any vessel in the King's service" and feared that "from the number of Rebel Ships now on this Coast, that not only the Albany but every Soldier on the Command [will fall] a Sacrifice."*

As a result of his efforts assisted by the pleas of General McLean, the sloops-of-war the HMS *North* and *Nautilus* remained at Penobscot, giving Mowat a total of three warships equipped with forty-four guns at his disposal.

Mowat's assessment of the *Albany* was probably correct as in 1782 it was declared unfit for combat service and converted to prison ship.

Accompanying the military force was the civilian guides and advisors who helped bring this expedition to fruition.

Dr. John Calef also joined the expedition as both the expedition's surgeon and acting chaplain. While some of the population fled upon the arrival of the British force, many remained cautiously neutral.

HMS Albany

John Nutting who had often surveyed his lands along the Penobscot served initially as the expedition's pilot, guiding the ships up the river. Upon arrival, General McLean appointed Nutting as the Overseer of Works co-coordinating the efforts of the local population that chose to help McLean's force.

Expedition's Surgeon/Chaplin

Loyalist DR. JOHN CIALEF,

His name was also spelled Calfe, Calf, Caleff and Kaloph, was born in Ipswich, Mass, Aug. 30, 1726, the son of Robert and Margaret (Stamford) Calef. His grandparents were Dr. Joseph and Mary (Ayer) Calef of Ipswich, who were married in Boston, May 2, 1693. The parents of Dr. Joseph were Robert arid Mary Calef of Roxbury, Mass. Robert Calef was the author of " More Wonders of the Invisible World," which antagonized Cotton and Increase Mather, about 1692. It was publicly burned on the campus of Harvard College by the orders of the latter, who was then president of the college.

Dr. John Calef had taken part in the siege of Louisburg and it is said he left a manuscript account of that event which has been lost. He was an important man in his time.

In 1755, the Governor ordered Dr. Calef to Fort Halifax, on the Kennebec River, now in the town of

Fort Halifax

Winslow, Maine, to attend the sick. He found his services much needed by the garrison and remained about two months. He also went there again in 1772. He engaged, as surgeon, in Colonel Ichabod Plaisted s regiment Feb. 18, 1756, to go to Crown Point, and was discharged Jan. 19, 1757, remaining at the Albany hospital.

He served in the Massachusetts General Court before the Revolutionary War, but remained loyal to the King and became obnoxious to the Colonists. He was declared by them a traitor, and a price was set upon his head. His wife helped him escape capture and he went to St. Andrews, N. B.

Agent of the Penobscot Loyalists

Dr. Calef was active in the Penobscot expedition and was Commissary of the inhabitants in the County of Lincoln, Maine. He was the surgeon at Fort George and acted as chaplain. In 1780, he went to England as an agent of the Penobscot Loyalists.

The scheme was that the country between the Saco and the St. Croix Rivers was to be erected into a new province to be called " New Ireland." Thomas Oliver was to be governor, Daniel Leonard, chief justice, and Dr. Calef the clerk of the council at a salary of 50.

The land was to be granted to the Loyalists in large tracts to the most meritorious with small grants to the poorest. It was to be a landed country. The English church was to be the established religion. This plan was approved of by the King.

Dr. Calef remained in England two years, when he revived the effort, but it received its deathblow from a decision of the Attorney General of England that it violated the sacredness of the chartered rights of the Province of Massachusetts Bay, and he was informed that it could not be done as " the pressure is too strong."

In 1784, Dr. Calef was one of the grantees of St. Andrews, N. B., and was the first physician to

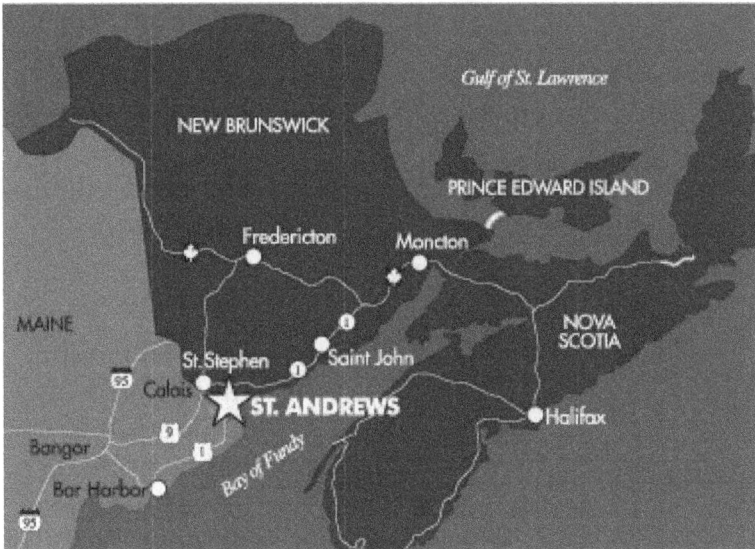

Location of St. Andrews, New Brunswick

settle there, where he built a house. After the war he was surgeon of the general hospital at St. John, N. B., and was attached to the garrison, then stationed at Fort Howe. After October, 1800, he returned to St. Andrews where he resided until his death, which occurred Oct. 23, 1812, at the age of 86 years.

Dr. Calef married, first, Margaret Rogers, daughter of Rev. Nathaniel and Mary (Leverett) Rogers of Ipswich, who died March 27, 1751 ; second, Jan. 18, 1753, Dorothy Jewett, daughter of Rev. Jedidiah and Elizabeth (Dummer) Jewett, both of Rowley, Mass.

27

Overseer of Works

JOHN NUTTING

Born in Corinth Center, Vermont, USA in 1745 to John Nutting and Sarah Chandler. John first married Mary Adams. He later married Hanna Fowler and had 10 children. He passed away on 1834 in Corinth Center, Vermont, USA.

Loyalist John Nutting

John Nutting, a Loyalist who had piloted Sir George Collier's expedition against Machias.

Lord Germain dispatched him to London to explain the situation in person. Nutting described the Castine peninsula as having a harbor that "could hold the entire British Navy" and was so easily defendable that "1,000 men and two ships" could protect it against any Continental force.

He also proposed that the strategic location of such a post would help carry the war to New England as well as offer protection for Nova Scotia from attack. Although Admiral Collier asked Nutting a year later what could possibly have induced him to recommend a settlement there (which he then denied). In light of subsequent events, Nutting's figures for defense at Castine proved to be amazingly accurate.

The Troops

William Hutchings, the "Last surviving Revolutionary War pensioner in New England" described the British soldiers who conducted the initial landing on the peninsula *"as frightened as a flock of sheep, and [they] kept looking around them as if they expected to be fired on by an enemy hid behind the trees."*

General McLean's force from the 74th and 82nd Foot consisted largely of untested soldiers recruited from Scotland the previous year. The 74th Foot, the Argyle Highlanders, formed in May of 1778 was recruited predominately from the Argyleshire region of Scotland.

This would be the first action for the two regiments. Although inexperienced, its soldiers had superior training and equipment when compared to their American counter-parts.

The local militia was almost destitute of arms and ammunition and failed to put up any semblance of a defense. McLean immediately made a conciliatory offer to the local population, promising to *"extend our protection, and give every encouragement, to all persons whatever denomination, without any retrospect to their former behavior, who shall, within eight days from the date hereof, take the oaths of allegiance and fidelity to his Majesty."*

Approximately four hundred and eighty inhabitants trekked to the British encampment to swear an oath. Approximately one hundred of them assisted in clearing the woods in front of the fort.

Factors that Influenced New England's Ability to Wage a Military Campaign

The Evolution of the Patriot's Militia System

In order to understand the ability of Massachusetts to conduct offensive operations, it is necessary to understand the militia system that produced the soldiers and leaders of the provincial armies.

Defensive in nature and local in its organization, the militia's narrow focus proved a constraint to the creation and implementation of provincial armies. Yet it served as the primary source of military manpower to the colony. Shaped by Puritan ideals, the militia compromised effectiveness to ensure local control and to prevent the governmental oppression observed in England.

Militiaman

A modification of the English militia system, fear of standing armies, and financial need constrained the Massachusetts system. The English militia was a two-tiered system consisting of: 1) the defensive trainbands (the name for a 17th or 18th century militia company in England and America) and 2) an untrained general militia, used for offensive operations.

32

Patriot Militia Were Primarily Defensive in Focus

With pressing defensive needs and a disdain for an offensive forces, the bay colony organized all men from sixteen to sixty exclusively into local trainbands (a division of civilian soldiers).

With limited means at the colony level, responsibility for equipping the militia fell on the individual and the town. Training for the militia remained limited and individually provided weaponry lacked standardization.

The establishment of provincial armies beginning with King Philip's War (1675-1676) stressed the militia system. The militia became a recruiting pool for provincial forces in addition to its local defense role. The colony established

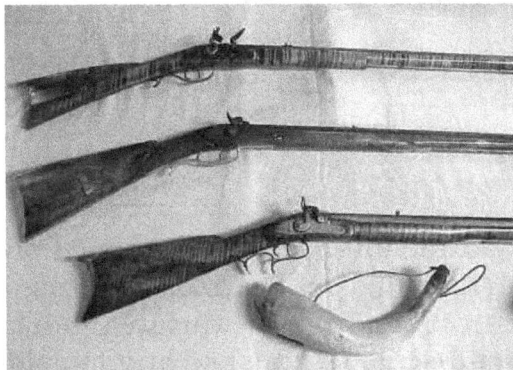

Revolutionary war weapons

locally based quotas and empowered the towns to impress soldiers when necessary. Impressment (the act of taking men into a military or naval force by compulsion, with or without notice) brought challenges to the town that struggled to balance local needs with the needs of the colony.

After experimenting with impressment during King Philips War, the colony focused on enticing volunteers instead through offering increasingly lucrative bounties and land grants.

33

Volunteer soldiers, while fitting within the democratic ideals of the colony, could be problematic. Familiarity between the soldiers and their popularly elected officers often led to slack discipline, especially when troops entered into large provincial armies.

Finally, a locally based militia system was not conducive to developing the leaders needed to lead expeditions and conduct larger-scale European-like operations such as the Penobscot Expedition.

Militia training

Wartime militia tasks consisted of town watch, garrisoning fortified blockhouses, or limited patrolling. Frontier warfare often consisted of small-scale skirmishes and lop sided ambushes rather than large standing battles. For successful frontier militia leaders, large-scale conventional operations were unfamiliar. Although regimental organizations existed, they were largely administrative. All of these issues made the transition from militia service to provincial duty problematic.

Militia System Evolves

Shortly after the Massachusetts Bay Colony established its charter; its General Court established a new militia system. Initial militia obligations, published 22 March 1631, charged the towns *"to enforce ownership of arms for all men and their servants and when necessary, provide arms on credit to those who could not afford them."*

"That [every] towne within this pattent shall, before the 5th of Aprill nexte, take espetiall care that [every person] within their towne, (except magistrates & ministers,) as well servts as others, furnished with good & sufficient armes allowable by the capt or other officers, those that want & are of abilitie to buy them themselves, others that are unable to have them pvided by the towne, for the present, & after to receive satisfaccon for that they disburse when they shal be able."

A few weeks later, the court expanded on the initial law by establishing a basic load of ammunition at one pound of powder, twenty rounds of shot, and two fathom of match enforceable by fines and mandated weekly drill. The colony soon realized that weekly drill was too demanding and by that November reduced them to a monthly basis. It revised the requirement to exclude

Colonial powder horns and shot flasks

the agricultural months of July and August. As a compromise to militia commanders, the General Court granted the authority to train unskilled men up to three days a week.

The town was the focal point of the militia system. Each town of sufficient size formed a company of militia. Initially company size was non-standard, but sixty-four men companies became the norm by 1672. If unable to field a company, towns nominated lesser officers to lead the militia detachment or combined with other towns to create a full company. Captains commanded companies assisted by a lieutenant, an ensign, three sergeants, and three corporals. Initially, only two-thirds of the men bore muskets with the remaining carried pikes.

Colonial Militia

Town Militias Organized into Three Statewide Regiments

In a series of militia laws, the colony aligned the companies into three geographically aligned regiments. The governor retained authority as the chief general of the militia. The first step towards a unified and standardized militia structure, the establishment of regiments set the groundwork for better coordination between militia companies.

The laws also established a system of popular election for officers. While in reality units only nominated their officers for the General Court's approval, but rarely did the court overturn election results.

American Revolutionary Soldiers

Local influence remained paramount, as the town's militia committee nominated the officers. Additional laws expanded the regimental structure, appointing sergeant majors to act as operational leaders.

The size of regiments, now aligned along county lines, made mustering the entire unit difficult, so the requirement dropped to once every three years. The regiment's officers met once or twice a year to discuss regimental business and coordinate defensive efforts. Largely administrative in nature, the regiment played little role in the tactical employment of the militia.

King William's War (1689 to 1697) and Expeditionary Experimentation

King William's War was a result of conflicting English and French expansionism and required more European style military operations. King William's War tested Massachusetts' ability to prosecute offensive expeditions.

French and Indian frontier warfare

Brutal raids and subsequent retaliation conducted by both the English and the French aided by Indian allies characterized frontier warfare with little discrimination between combatants and unarmed women and children. Communities along the frontier remained challenged by the need to provide for both its own defense and for the colony's operations.

Massachusetts' entry into the King William's war began the revocation of its charter by King Charles II in 1684 and its inclusion into the New England Dominion in 1686. The colonies, particularly Massachusetts, proved to be difficult to control and often resisted English policies. After a series of incidents, in 1684 Charles II revoked the charter of the Massachusetts Bay Colony.

Bagaduce Peninsula Raid Starts a French/Indian War – 1688

The royally appointed governor, Sir Edmund Andros, sought to strengthen English claims along the contested border between Massachusetts and Canada. He planned a series of outposts along the

St Castin's home destroyed by Andros

coast of Maine. After building a series of forts garrisoned by British troops brought from New York, he reached his most eastern point of his campaign at the Bagaduce Peninsula on the Penobscot Bay.

The Peninsula was the home of Frenchman Baron Jean-Vincent de St. Castin who established a trading post with the Indians. St. Castin had cultivated his relationship with the local Indian tribes and had even married the daughter of one of the local chiefs. Andros sought to convert St. Castin to the English side, but failing that destroyed his home.

The Indians considered the act an insult and launched a series of retaliation raids that would blossom into full-scale war. Meanwhile in England, the "Glorious Revolution" resulted in the assumption of William and Mary to the throne and spelled the end of the New England Dominion.

Puritan leaders quickly arrested Andros when he returned to Boston. Although stripped from power, the frontier war Andros helped create continued unabated. The French and their Indian allies took all of the new forts one by one.

This collapsed Massachusetts's eastern frontier and led to a significant depopulation of Maine.

British Royal troops abandoned their posts along the Maine coast and those that remained were often removed by Massachusetts for their real or suspected ties to the Anglican Church.

The following May a French and Indian war party from Quebec captured the last fort at Casco, Maine.

Baron Jean-Vincent de St. Castin

Massachusetts Gains Military Confidence - 1745

Fortified towns are hard nuts to crack, and your teeth are not accustomed to it. Taking strong places is a particular trade, which you have taken up without serving an apprenticeship to it. Armies and veterans need skillful engineers to direct them in their attack. Have you any? But some seem to think that forts are as easy taken as snuff.

— Benjamin Franklin, 1745

In the three decades prior to the American Revolution, Massachusetts continued to refine its ability to raise and support provincial armies. This era began with improbable success at Louisbourg, Nova Scotia and ended with Massachusetts in a comfortable rhythm of generating provincial armies. Funding from Britain enabled the colony to generate largely volunteer armies, targeting a population of men described by historian Fred Anderson as *"temporarily available for military service"*.

While there was not a surplus of labor in Massachusetts, young men often chose military service as a way to earn a plot of land and what essentially amounted to startup money. While some raids occurred along the Massachusetts frontier, most of the militia's efforts during these wars were more precautionary than actual.

With limited home defense requirements, the militia could better manage the compromise between local and colony manpower requirements.

The last two wars with the French also employed a large percentage of the population at some time during the wars. It shaped a new generation of leaders for the colony. Men who served at Louisbourg led the colony during the Seven Years War. Similarly, that war provided many of the leaders of the American Revolution.

The landing of troops from New England on the island of Cape Breton to attack Louisbourg - 1745

The decisive battle of King George's War, Massachusetts' triumph at Louisbourg was a product of many factors. Volunteerism was high, fueled by religious fervor, and local defense needs were minimal. The contribution by the Royal Navy, although often downplayed, was significant by both isolating Louisbourg and capturing critical supplies.

William Pepperell provided solid leadership and in some cases, his subordinate leaders even demonstrated initiative. Finally, the expedition had some outright lucky breaks that overcame some deficiencies of the army. For Massachusetts, the

Yankee spirit and ingenuity allowed an inexperienced army of farmers and fishermen to defeat well- entrenched regular troops.

After the first two years of the Seven Years War, the role of Massachusetts's provincial armies declined to a supporting role to the British regulars. Providing armies on a relatively predicable schedule, Massachusetts became fairly proficient at assembling them.

The siege of Louisbourg was New England's most successful military expedition conducted to date. Yet as spectacular as the victory was, it led Massachusetts to some dangerous conclusions on the capability of her militia.

Conducted akin to a business transaction, enlistment contracts specified the duration, pay, and even the army's objective specified as part of the contract. Despite drawing criticism from British officers for its cost and inefficiency, the financial benefits of the provincial army enlistment system suited Massachusetts and its men well.

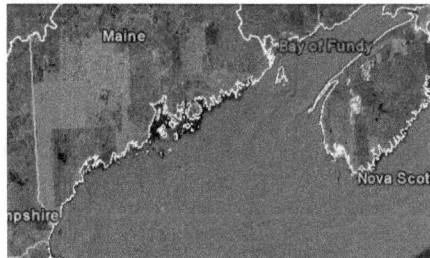

As the revolution flared in Boston, Maine initially was of little military interest to the British and many of its residents hoped they would gain some degree of protection by the province's insignificance. Instead, Maine found itself once again the frontier between Massachusetts and British held Nova Scotia.

Revolutionary incidents in Falmouth and Machais drew the attention of the Royal Navy and compounded by rampant privateer attacks off the Maine coast.

Colonist Supply the Men, British the Leadership

The Seven Years War marked the introduction of British control in military affairs. Aside from the opening campaigns, Massachusetts took on the role as a force provider instead of a director of the war. For the militia, the demands of the war streamlined the system the colony used to generate its provincial armies. Massachusetts used crown funding to gain volunteers through enlistment bounties. Yet reliance on volunteers made its armies more subject to the demands of the individual soldier.

By using the support of the Royal Navy, Massachusetts proved it could plan, equip, and lead a large conventional campaign. Massachusetts now possessed the manpower and resources to generate and field an army. Yet it also marked the beginning of increased British involvement and management of the English colonies.

By 1779, Royal Navy raids on the Maine coast posed challenges for the militia. The Colonial militia once again had to balance local defense needs with the needs of the colony.

In order to convince a man to volunteer, the colony had to provide financial incentives and set terms of enlistments. The result was armies and expeditions that came with their own limitations of set enlistment timelines and when possible financial bounties. Although not efficient or ideal from the view of the colony or a military commander, it allowed the colony to generate forces it needed from the militia that could expand or contract as the conditions changed.

The Penobscot Expedition Saga Begins

In June 1779, while most of the Revolutionary War was focused in the southern United States, a small British fleet landed two regiments of 700 soldiers on the Castine Peninsula, in the upper reaches of Penobscot Bay, Maine. Intent on establishing a base from which they could operate more effectively against American privateers, ensure the extraction of valuable naval stores, and develop a refuge for displaced Loyalists, the British enlisted local support and began the construction of Fort George.

Word reached Boston quickly (Maine was a district of Massachusetts until 1820), and over the ensuing month the largest American naval force of the Revolutionary War, known as the Penobscot Expedition, was assembled.

British Land at Castine

On June 17 1779, British Army forces under the command of General Francis McLean landed and began to establish a series of fortifications centered on Fort George, located on the Majabigwaduce Peninsula in the upper Penobscot Bay, with the goals of establishing a military presence on that part of the coast and establishing the colony of New Ireland.

British forces arrive

Nutting reached New York in January 1779, but General Clinton had received copies of the orders from other messengers. Clinton had already assigned the expedition to General Francis McLean who was based in Halifax and thus sent Nutting there with Germain's detailed instructions.

McLean's expedition lands at Castine

McLean's expedition set sail from Halifax on May 30, 1779, and arrived in the Penobscot Bay on June 12. The next day McLean and Captain Andrew Barkley, the commander of the naval convoy, identified a suitable site at which they could establish a post.

On June 16, his forces began landing on a peninsula that was then called Majabigwaduce (now Castine), between the mouth of the Bagaduce River and a finger of the bay leading to the Penobscot River.

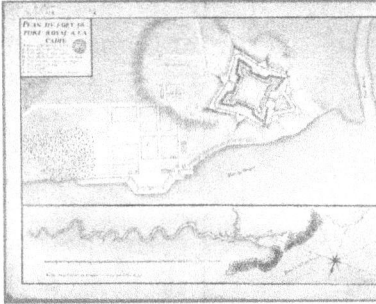

Plan for Ft George

The troops numbered approximately 700: 50 men of the Royal Artillery and Engineers, 450 of the 74th Regiment of (Highland) Foot and 200 of the 82nd (Duke of Hamilton's) Regiment. These began to build a fortification on the peninsula, which jutted into the bay and commanded the principal passage into the inner harbor.

Fort George was established in the center of the small peninsula with two batteries outside the fort to provide cover for the Albany, which was the only ship expected to stay in the area. Construction of the works occupied the troops for the next month, until rumors came that an American expedition was being raised in Boston to oppose them, following which efforts were redoubled to have works suitable for defense against the Americans prepared before they arrived.

Captain Henry Mowat of the Albany, who was familiar with Massachusetts's politics, took the rumors (which were followed by reports that a fleet had left Boston) quite seriously, and convinced General McLean to leave additional ships that had been part of the initial convoy as further defense. Some of the convoy ships had already left; orders for armed sloops *North* and *Nautilus* were countermanded before they were able to leave.

Massachusetts Reacts

Within days, letters bearing the news of the invasion began pouring into the General Court in Boston. A letter from Brigadier General Charles Cushing, the commander of the militia of eastern-most county, Lincoln County, which encompassed the Penobscot River requested immediate support and supplies.

"There is a great difficulty in the way of the Militia of this County turning out not having provisions to support them, not one Family in Ten having Bread in their Houses nor anything else scarcely except from day to day. Neither is there a sufficiency of arms nor ammunition."

He further recommended that a joint naval and land attack could easily dislodge the British outpost. With little deliberation, on 24 June the General Court ordered the Board of War prepare all state and private armed vessels that can be readied within six days to sail. As an incentive to private ship owners, the state promised that *"in case the said vessels or any of them shall be lost or damaged while on said Expedition, this State will make good such loss or Damage."* Furthermore *"The Officers and Seamen of such armed Vessels [are allowed] the same Pay, Rations and Privileges allowed to the Continental Navy."*

To encourage recruitment and to maintain secrecy, Massachusetts also imposed a twenty-day embargo on all non-fishing related, outward bound ships. Four days later, as the enormity of the task became more apparent, the General Court resolved to seek the services of Continental ships at Boston Harbor, and granted authority to impress any private ships required for the expedition.

To aid in this endeavor, Massachusetts later extended the embargo and additional twenty days further attempting to coerce non-participating ships.

In reality, assembling a fleet for the expedition would take the better part of a month. Ultimately, Massachusetts assembled nineteen armed ships for the expedition.

Boston Harbor circa 1779

Three Continental Navy vessels anchored in Boston when news arrived. Massachusetts gained approval from the Navy Board to use these three ships, the *Warren, Province,* and *Diligent.* Designated the commander of the fleet was the senior Continental Naval Officer Commodore Saltonstal commander of the *Warren.* All three ships were short crewmen and Massachusetts endeavored to fill the shortages.

The new thirty-two-gun frigate and the fleet's flagship, the *Warren* alone required one hundred seamen. From the Massachusetts Navy came the *Tyrannicide, Hazard,* and *Active.*

New Hampshire contracted the privateer the *Hampden* to contribute to the expedition. Massachusetts chartered the remaining twelve privateers giving the fleet a total of three hundred and forty-four guns. Massachusetts contracted or impressed twenty-four transports to transport the supplies and soldiers to the Penobscot Bay.

Assigned to the continental and state ships were approximately three-hundred marines who would prove their worth in the coming battle.

Tyrannicide, Hazard, and *Active*

The American Command is Assembled

Commodore Dudley Saltonstall (1738–1796)

Early life - Dudley Saltonstall was born in 1738 to Gurdon Saltonstall Jr and Mary Winthrop. Both sides of his family were prominent in British colonial politics; his great-grandfather on his father's side was Sir Richard Saltonstall, and his mother was descended from John Winthrop, who served as governor of the Province of Massachusetts Bay in the 17th century.

His father was a prominent figure in New London and Connecticut politics, serving as a probate judge and a leader of the community. In 1765 he married Frances Babcock, the daughter of Joshua Babcock, a doctor and lawyer who served on the supreme court of the Rhode Island Colony.

Saltonstall took positions on the ships of the colonial mercantile fleet, and served as a merchant captain during the Seven Years' War. In April 1762 he was given command of a letter of marque brigantine, the *Britannia*, with which he made several successful voyages to the West Indies. During these years he established a reputation as a competent ship's captain.

Joins Continental Navy

When the American Revolutionary War broke out, Saltonstall joined Connecticut's militia, helping to defend New London's harbor. When the Continental Navy was established, he was given one of the first captain's commissions, based on the recommendation of his brother-in-law Silas Deane, who served on Connecticut's Naval Committee. He was given command of the *Alfred*, the flagship of the new navy.

He hired John Paul Jones as his first lieutenant, and gave him the responsibility of overseeing the fitting out of the newly acquired ship. He and Jones did not get along well, as Jones did not like Saltonstall's sometimes distant and superior demeanor.

Saltonstall captained the *Alfred* on the Continental Navy's maiden voyage in March 1776, an expedition to Nassau in the Bahamas whose objective was arms and critically needed gunpowder. The expedition was somewhat successful, as Nassau was taken, but its governor had managed to remove much of the gunpowder before the takeover was completed.

HMS Glasgow

The return voyage was uneventful, although smallpox was spreading through the ships' crews, until the fleet neared Block Island. On April 4 and 5, the fleet captured British ships. On the morning of April 6, the *HMS Glasgow* was spotted, and was brought to action. In the ensuing battle, *Alfred's* steering controls were damaged by cannon fire, and she drifted out of the action. The *Glasgow* was able to escape to Newport outrunning the heavily laden fleet.

While the expedition was successful, commodore Hopkins and Saltonstall were questioned over the fleet's failure to follow its stated orders, which had been to engage the British fleet off the Carolinas, and over its failure to capture the clearly

USS Trumbull

outnumbered *Glasgow.* Saltonstall was not censured, but other captains in the fleet were punished for cowardly behavior.

In September 1776 Saltonstall was given command of the USS *Trumbull,* which had been built at a dockyard on the Connecticut River. To Saltonstall's dismay, she was too heavily laden to cross the sand bar at the mouth of the river.

After numerous repeated attempts to pass over the bar in 1776 and 1777, the Marine Committee gave her command to Elisha Hinman, who successfully floated her into Long Island Sound in August 1779.

Saltonstall's next command was the *Warren,* based in Boston. He replaced commodore Hopkins at her helm in July 1779; Hopkins was suspended for breach of orders, and was eventually dismissed from the navy.

Given Command of the Penobscot Expedition

In the summer of 1779, the state of Massachusetts (which at that time included the District of Maine), organized an expedition to dislodge the British from this position. Saltonstall, the senior Continental Navy commander, was given command

USS Warren

of the naval forces, which consisted primarily of ships from the Massachusetts State Navy, a large number of privateers, and a few Continental Navy ships, include the *Warren*. Command of the land forces accompanying the expedition was given to a relatively inexperienced Massachusetts militia brigadier general, Solomon Lovell.

Summary of his Penobscot Command

As the senior Continental Navy officer, command of the Penobscot Expedition fleet defaulted to Captain Dudley Saltonstall. Known as snobbish and ill tempered, Saltonstall gained his commission in the Navy due to the influence of his brother-in-law Silas

While he proved able enough during single-ship combat, he had not commanded anything similar in size to this fleet. His instructions from the navy Board charged him to *"Take every Measure & use your Utmost Endeavous [sic] to Captivate, Kill, or destroy the Enemies whole Force both by Sea & land & the more effectually to answer that purpose you are to Consult Measures & preserve the greatest harmony with the Commander of the Land Forces that the Navy & Army may cooperate & assist each other."*

Later career

Saltonstall returned to Connecticut, and convinced one of his wife's relatives, Adam Babcock, to support him in a privateering venture. As captain of the 16-gun brig *Minerva*, he embarked on a successful career as a privateer in 1781. Among his prizes; the richest captured by a Connecticut ship; the British ship *Hannah* was valued at £80,000.

After the war, he engaged in trade with the West Indies, and also dabbled in the slave trade. He died in 1796 in the West Indies, apparently of a tropical disease.

Second in Authority

Brigadier-General Peleg Wadsworth 1748 - 1829

He served as an aide under General Artemas Ward in 1776 and was later present at the Battle of Long Island. He received an appointment as a brigadier in the militia in 1777 and promoted to the position as the state adjutant general the following year. Younger and full of energy, he complimented the older General Lovell.

Brigadier General Peleg Wadsworth

Peleg Wadsworth was an American officer during the American Revolutionary War and a Congressman from Massachusetts representing the District of Maine. He was also grandfather of noted American poet Henry Wadsworth Longfellow.[1] Wadsworth was born in Duxbury, Massachusetts, to Peleg and Susanna (Sampson) Wadsworth. He graduated from Harvard College with an A.B. (1769) and an A.M. (1772), and taught school for several years in Plymouth, Massachusetts, with his former classmate Alexander Scammel. There he met Elizabeth Bartlett (1753 to 1825), whom he married in 1772.

The Wadsworths lived in Kingston, Massachusetts, until 1775, when Wadsworth recruited a company of minutemen, of which he was chosen captain. His company mustered in response to the alarms generated by the Battle of Lexington and Concord on April 19, 1775. The Plymouth County battalion, commanded by Col. Theophilus Cotton marched to Marshfield, Massachusetts to attack a garrison of

British troops there. The attack was delayed for two days, allowing the British time to escape Marshfield by sea. During that time, Capt. Wadsworth, frustrated with the delay, advanced his company to within firing range of the British encampment, nearly instigating combat.

Wadsworth served as aide to Gen. Artemas Ward in March 1776, and as an engineer under Gen. John Thomas in 1776, assisting in laying out the defenses of Roxbury, Massachusetts. He was present at the Battle of Long Island on August 1, 1776. He was made brigadier general of militia in 1777 and Adjutant General of Massachusetts in 1778.

Fort George and Dungeon, Built by British in 1779.

Wadsworth escapes the British from Castine

In March 1780, Peleg was given command of all the troops raised for the defense of the Province of Maine. On February 17, 1781, British soldiers overran his headquarters in Thomaston. Wadsworth was captured and imprisoned in Fort George at Bagaduce (Castine) (the same fort he had led the attack against in the summer of 1779), but he and fellow prisoner Maj. Benjamin Burton eventually escaped by cutting a hole in the ceiling of their jail and crawling out along the joists. Wadsworth then returned to his family in Plymouth, where he remained until the war's end.

From a letter dated June 22, 1781 to Commander Sir Henry Clinton from Col Campbell Cmdr at Ft George in Castine:

"I am sorry to inform your Excellency that on the

The Wadsworth-Longfellow House made famous by his grandson Henry Wadsworth Longfellow.

19th instant, the Rebel Brigadier General Wadsworth who was confined here in a Barrack Room, With one Burton once a Major of Militia, made their escape, by cutting a Hole in the ceiling of the Room, altho two Sentrys were constantly Posted at the Door of the Room and a Window but in the Door for the Sentrys to look through to observe their Motions."

In April 1784 Wadsworth returned to Maine, purchased 1.5 acres of land on Back Street (now Congress Street in Portland), engaged in surveying, and opened a store in early 1785. There he also built a house, now the historic Wadsworth-Longfellow House.

He headed the committee that organized the first convention to discuss independence for Maine from Massachusetts, held in January 1786. He and his wife had eleven children.

Although he continued to live in Portland, in 1790 he purchased 7,800 acres from the Commonwealth in what became the town of Hiram, Maine, settled

Wadsworth family cemetery in Hiram, Maine

his son Charles there in 1795, and also in 1795 began building Wadsworth Hall for his retirement.

In 1792 Wadsworth was chosen a presidential elector and a member of the Massachusetts Senate, and from 1793-1807 was the first representative in Congress from the region of Massachusetts that later became Maine. In January 1807 he moved to Hiram where he incorporated the township (February 27, 1807) and served as selectman, treasurer and magistrate. For the remainder of his life he devoted himself to farming and local concerns. He died in Hiram on November 12, 1829, and is buried in the family cemetery at Wadsworth Hall.

Wadsworth's Portland house was declared a National Historic Landmark for its association with him and with his grandson. Wadsworth Hall, his home in Hiram, is listed on the National Register of Historic Places.

Command of the Ground Forces

Brigadier General Solomon Lovell 1732 - 1801

Solomon Lovell was born in Abington, in the Province of Massachusetts Bay, on June 1, 1732, to David and Mary (née Torrey) Lovell. His father was a Harvard graduate, teacher, and sometime preacher. He died when Solomon was quite young, and the boy was raised first by his grandfather Enoch Lovell, and after his death by his stepfather, Samuel Kingman. Kingman, a military man, may have influenced the young Solomon to develop an interest in the military. Lovell's military service during the French and Indian War (1754–1760) is not known in detail; he is known to have served as a first lieutenant in a militia company at Lake George, New York during the 1756 campaign. In 1758 he married Lydia Holbrook. The couple had two children; the first died in infancy, and Lydia died during the birth of the second in 1761. The following year Lovell remarried, to Hannah Pittey, a woman who had originally spurned his proposal to her before his first marriage. With Hannah he settled into her house in Weymouth; they had seven children, three of whom survived to adulthood. He was active in town affairs, and began serving in the provincial assembly in 1771. He was also active in the local militia, rising to the rank of major in July 1771 and colonel in 1775.

Brigadier General Solomon Lovell

61

With the outbreak of the American Revolutionary War with the Battles of Lexington and Concord in April 1775, Lovell's military activity increased. He was commissioned a colonel of the 2nd Massachusetts Regiment in February 1776, and his troops were among those that occupied Dorchester Heights, south of Boston, prompting the British to withdraw the city. He continued to be

Lovell at Dorchester heights and the British Evacuation of Boston

active in the defense of eastern Massachusetts, and was promoted to brigadier general of the Suffolk County militia on June 24, 1777. Lovell led Massachusetts's troops in the 1778 Battle of Rhode Island, where Lovell was one of several officers who "distinguished themselves by their coolness and courage."

Lovell had, during the war, periodically served as a representative to the state legislature. He continued to do so after the war, also occasionally serving as town selectman. When Norfolk County was separated from Suffolk County, Lovell was given the task of petitioning the legislature to keep Weymouth a part of Suffolk County. He was unsuccessful in this effort; Weymouth is now in Norfolk County.

He first saw action in 1756 at Crown Point during the French and Indian War. Lovell later saw service both during the revolution at the siege of Boston and in Rhode Island in 1778 where he commanded nearly 1,200 men from Massachusetts.

As part of General John Sullivan's army, his force was part of a portion of the army commanded by General Nathaniel Greene. His unit served with

Lovell during the French and Indian War

distinction, receiving recognition from Sullivan *"for their intrepidity, which they showed in repeatedly repulsing the enemy, and finally driving them from the field of action."*

Furthermore, he was a respected member of the General Court, representing the town of Weymouth since 1771.

By 1779, he developed a reputation as a popular, steady, and capable officer. By all measures, he appeared to be the perfect successor to the legendary William Pepperell who successfully took Louisbourg decades earlier.

Lovell died in Weymouth on September 9, 1801, having outlived his wife by six years. He is buried in the Pittey family tomb in Weymouth.

Lieutenant Colonel Paul Revere 1735 - 1818

Lieutenant Colonel Paul Revere

He was appointed as Lovell's chief of ordinance. No doubt an ardent patriot, Revere had little military experience, the majority of that time spent in garrison duty on Castle Island. His appointment owed much to his political connections that earned him many enemies.

Despite his undoubtedly patriotic spirit, Revere showed little energy and vigor during the campaign.

Paul Revere was a silversmith and patriot from Massachusetts during the American Revolution.

He is most famous for alerting local militia of the approaching British forces shortly before the battle of Lexington and Concord.

The following are some facts about Paul Revere:

Born in the North End of Boston in December of 1734, Revere's father was Apollos Rivoire, a French Huguenot immigrant who later changed his name to Paul Revere to fit in with the English immigrants in the city. Revere's mother was Deborah Hichborn, a daughter of a local artisan family.

Paul Revere served as an apprentice in his father's goldsmith shop. After his father died when Paul was 19 years old, he took over his father's shop and became responsible for his large family.

Paul Revere in the French and Indian War

At age 21, Paul Revere volunteered to fight in the French and Indian War at Lake George in New York and was appointed second lieutenant in the colonial artillery.

A portrait of Paul Revere by by John Singleton Cople circa 1769

Revere only served a short stint in the war during a failed exhibition to Lake George and returned to Boston without seeing much military action.

He soon returned to civilian life and married Sarah Orne in 1757. Together they had eight children. Three of Revere's daughters later married into Abraham Lincoln's family and three of his grandsons fought in the Civil War.

Paul Revere in the American Revolution

Although Revere's silversmith shop was successful and his work was sought after, the economic depression before the American Revolution hit his business hard and he was forced to supplement his income by working as an engraver, a courier and also as a dentist (installing false teeth with metal wires.)

An important step in Revere's life was when he joined the Masonic Lodge of St. Andrew in September of 1760. Here he met patriot activists such as Joseph Warren, James Otis and John Hancock and soon became involved in the activities of the American Revolution.

Paul Revere and the Sons of Liberty

Around the same time that Revere joined St. Andrew's lodge, he also joined the Sons of Liberty, a group of political militants who organized protests against British forces.

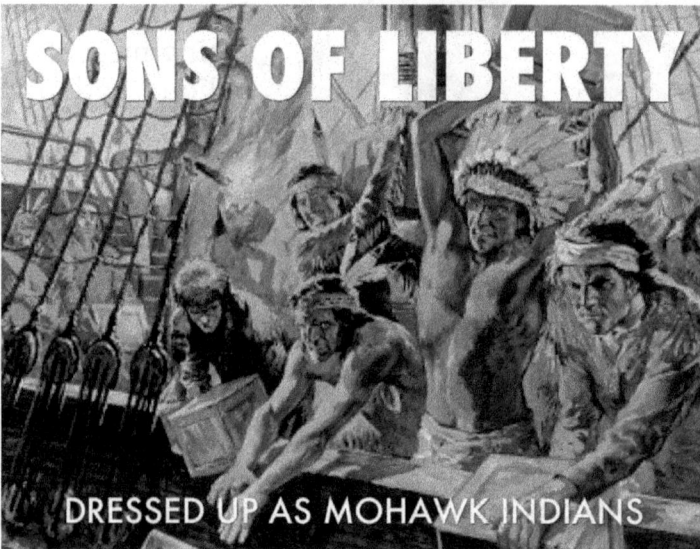

SONS OF LIBERTY

DRESSED UP AS MOHAWK INDIANS

The Sons of Liberty, who used the Green Dragon Tavern as their headquarters, were responsible for dumping millions of dollars worth of tea into Boston harbor during the Boston Tea Party, which Paul Revere took part in.

Tragedy struck when Revere's wife died in childbirth in 1773, leaving him a widower with a newborn and many children to care for. He remarried later in the year to a woman named Rachel Walker, with whom he had eight children.

In the fall of 1774, Revere founded one of the first spy rings in America, the Mechanics, to keep track of British troop movements, according to the book Paul Revere's Ride:

"Many years later he [Revere] recalled that 'in the Fall of 1774 and Winter of 1775, I was one of upwards of thirty, chiefly mechanics, who formed ourselves into a committee for the purpose of watching the movements of the British soldiers, and gaining every intelligence of the movements of the Tories. We held our meetings at the Green Dragon Tavern.'"

The Mechanics were eventually infiltrated by a British spy working for General Thomas Gage. Although Revere never discovered the identity of the spy at the time, it was later revealed to be Dr. Benjamin Church.

Unlike many other patriot activists at the time, such as Samuel Adams and John Hancock, Revere was not a member of the noble class and aside from his activities in the mason lodge, his limited education and vocation as an artisan prevented him from traveling in the same social circles as many of the other activists.

The Midnight Ride of Paul Revere
It was Revere's side job as a courier for the Boston Committee of Public Safety and his involvement in the mason lodge that led to his famous ride.

On the night of April 18, 1775, fellow lodge member Dr. Joseph Warren instructed Revere and William Dawes to ride to Lexington and warn John Hancock, Samuel Adams and local militia of approaching British forces.

Revere and Dawes met local physician Samuel Prescott, who decided to join them. The ride was later immortalized in a poem by Henry Wadsworth Longfellow titled "Paul Revere's Ride."

Revere also wrote his own account of his famous ride and his eventual capture by British troops:

"When we had got about half way from Lexington to Concord, the other two stopped at a house to awake the men, I kept along. When I had got about 200 yards ahead of them, I saw two officers as before. I called to my company to come up, saying here was two of them, (for I had told them what Mr. Devens told me, and of my being stopped). In an instant I

Statue of Paul Revere near the Old North Church in Boston

saw four of them, who rode up to me with their pistols in their bands, said "G—d d—n you, stop. If you go an inch further, you are a dead man." Immediately Mr. Prescot came up.

We attempted to get through them, but they kept before us, and swore if we did not turn in to that pasture, they would blow our brains out, (they had placed themselves opposite to a pair of bars, and had taken the bars down). They forced us in. When we had got in, Mr. Prescot said "Put on!" He took to the left, I to the right towards a wood at the bottom of the pasture, intending, when I gained that, to jump my horse and run afoot.

68

Just as I reached it, out started six officers, seized my bridle, put their pistols to my breast, ordered me to dismount, which I did. One of them, who appeared to have the command there, and much of a gentleman, asked me where I came from; I told him. He asked what time I left. I told him, he seemed surprised, said "Sir, may I crave your name?"

I answered "My name is Revere. "What" said he, "Paul Revere"? I answered "Yes." The others abused much; but he told me not to be afraid, no one should

Paul Revere House, Boston, Mass

hurt me. I told him they would miss their aim. He said they should not, they were only waiting for some deserters they expected down the road. I told him I knew better, I knew what they were after; that I had alarmed the country all the way up, that their boats were caught aground, and I should have 500 men there soon. One of them said they had 1500 coming; he seemed surprised and rode off into the road, and informed them who took me, they came down immediately on a full gallop. One of them (whom I since learned was Major Mitchel of the 5th Reg.) clapped his pistol to my head, and said he was going to ask me some questions, and if I did not tell the truth, he would blow my brains out. I told him I esteemed myself a man of truth, that he had stopped *me on the highway, and made me a prisoner, I knew not by what right; I would tell him the truth; I was not afraid."*

After the Revolutionary War began, Revere served as a lieutenant colonel in the Massachusetts State Train of Artillery and commanded Castle Island in the harbor.

A portrait of Paul Revere by Gilbert Stuart in 1813

Revere's military career was unremarkable and ended with the failed Penobscot expedition in 1779 during which he disobeyed orders and was charged with insubordination, ordered to resign command of Castle Island and was placed temporarily under house arrest.

Paul Revere After the American Revolution

After the revolution, Revere expanded his business and began exporting his goods to England. He also ran a small hardware store until 1789 and ran his own foundry where he made bolts, spikes and nails for local ships. Revere also produced cannons and cast bells.

In 1801, Revere opened the first copper rolling mill in America and created copper sheeting for the hull of the U.S.S. Constitution and the dome of the Massachusetts States House in 1803.

In 1804, Revere befriended Deborah Sampson, a woman who had disguised herself as a man and fought in the American Revolution, and was so impressed with her story that he wrote a letter to Congress asking them to award her a pension for her service. Sampson was awarded a pension the following year.

Revere buried in the Granary Burial Grounds

Revere continued to work well into his old age before he finally retired at the age of 76 in 1813. That same year, Revere became a widower again in 1813 when his wife Rachel died after a short illness.

Paul Revere in the War of 1812:

In September of 1814, Paul Revere volunteered his manual labor during the ongoing War of 1812, to help build Fort Strong on Noddle's Island to protect Boston from the threat of British invasion.

Paul Revere's Death and Burial

On May 10, 1818, Paul Revere died of natural causes at 83 years of age and was buried in Boston's Granary Burying Ground.

Surgeon General

Dr. Eliphalet Downer. (1744-1806) -

Eliphalet Downer

Volunteer Surgeon at Lexington and Concord, 19th April, 1775; Surgeon of Heath's Massachusets Regiment, May to December, 1775; Surgeon 24th Continental Infantry, 1st January to 31st December, 1776; served subsequently in United States Navy.

Dr. Eliphalet Downer had a hand-to-hand struggle with a redcoat during the British retreat from Lexington. "Landmarks of Middlesex." Dr. Downer married Mary Gardner, native of Brookline, and they had five children.

His son was one of the Founders of the Horticultural Society. One of his daughters married Mr. John Hancock, and survived her husband many years. She owned and lived in the cottage, which was recently taken down, just west of the house of the Good Shepherd.

Adjutant General

Capt. Jeremiah Hill of Biddeford, Maine.
1747-1820

Jeremiah Hill, (Jr.) born April 30, 1747; married Mary Emery (b. Mar.26, 1752) Sept. 6, 1772, daughter of Obed and Sarah (Dyer) Emery of Biddeford.

He was a captain in James Scammon's York County 30th regiment at Cambridge in 1775. Scammon's Regiment was raised in answer to act by Second Provincial Congress of Massachusetts

Battle for Bunker Hill, Charlestown, MA

on April 23, 1775, four days after the battles of Lexington and Concord. They marched to Cambridge on receipt of orders dated May 10th, 1775 and reported for duty on May 23, 1775.

During the battle of Bunker Hill, Scammon had marched from Cambridge to Lechmere's Point (East Cambridge) around two to observe the movement of the British floating batteries. Here Scammon was ordered by Col. Whitcomb to march to the hill, which Scammon interpreted to mean Cobble Hill (McLean Asylum), whence Scammon sent messangers to General Putnam for clarification. Before their return, Scammon hurried his troops forward to the fight. The delay in orders had Scammon's regiment reach no further than Bunker Hill, meet with retreating soldiers from Breed's Hill and then immediately join the retreating troops.

On July 13, a court martial was convened for disobedience of orders and for not showing proper spirit during the battle. He was acquitted of charges.

On January 1, 1776, was commissioned captain in Col. Edmund Phinney's 18th Continental Regiment. Stationed at Fort Independence and Fort George through 1776. On January 1, 1777, commissioned Captain in Col. Joseph Vose's 1st Massachusetts regiment, and resigned November 4, 1777. He joined the 1st Massachusetts regiment at West Point, and took part in the Saratoga campaign. He was commissary of prisoners in Rhode Island, in 1778, and was adjutant-general of the Bagaduce Expedition in 1779.

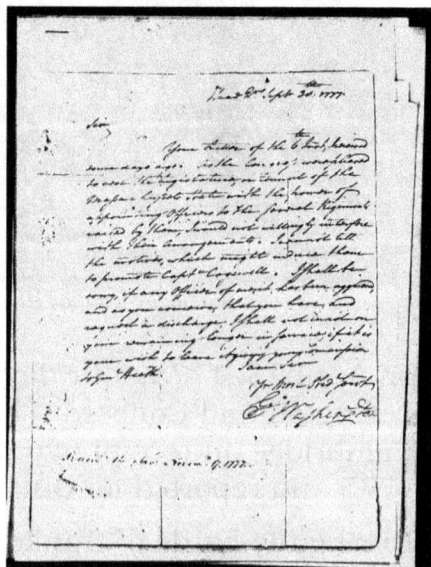

Letter from Geo. Washington to Capt. Hill

Hill wrote George Washington on 6 Sept.: *"I have inclosed the Copy of a Letter which I offer'd to the honorable Council and House of this State (Massachusetts) which was committed to the honorable House who thereupon voted to recommend me to be taken such Notice of by your Excellency as my Merit deserved or to discharge me with Honor from the Service if I desired it which Letter was sent to the Council who refused to concur with it; thrô the Influence of the honorable General Ward because he thought it did not do Justice to my Merit; I am much obliged to his Honor, Therefore Dear Sir I am constrain'd once more to request a Discharge"*

The enclosed letter complains of the promotion of Thomas Cogswell to major of the 1st Massachusetts Regiment ahead of Hill and includes a copy of a petition of 1 July 1777 that he had sent to GW.

Hill wrote that "Waiting sometime but receiving no Answer I waited on his Excellency in Person (my Petition has escaped his Excellency's particular Notice) his Excellency was pleased to decline giving me a Discharge but offerd me Liberty to come to this honorable Court for Redress"

George Washington responds:

Head Qrs [Skippack, Pa.] Septr 30th 1777.

Sir

Your Letter of the 6th Instant I received some days ago. As the Congress were pleased to vest the Legislature, or Council of the Massachusets State with the power of appointing Officers to the Several Regiments raised by them, I would not willingly interfere with their Arrangements. I cannot tell the motives which might induce them to promote Capn Cogswell. I shall be sorry if any Officer of merit has been aggrieved, and as you conceive, that you have, and request a discharge, I shall not insist on your remaining longer in service if it is your wish to leave it, giving your Commission to Genl Heath. I am Sir Yr Most Obedt servant

Go: W.

Jeremiah was tried for heresy in the Church of Christ, Biddeford, ME in 1793. There are transcripts in Devon, England.

Capt. Hill was a representative to the General Court, a justice of peace, and was the first collector of Saco, 1789 to 1809. He contracted smallpox at Boston in April, 1776, and died June 11, 1820, aged seventy-three years.

An Hiftorical   *1166. h. 9.*

ACCOUNT

OF THE

SMALL-POX

INOCULATED

I N

NEW ENGLAND,

Upon all Sorts of Perfons, *Whites, Blacks,* and of all Ages and Conftitutions.

With fome Account of the Nature of the Infection in the NATURAL and INOCULATED Way, and their different Effects on HUMAN BODIES.

With fome fhort DIRECTIONS to the UNEXPERIENCED in this Method of Practice.

Humbly dedicated to her Royal Highnefs the Princefs of W A L E S, by *Zabdiel Boylfton,* Phyfician.

L O N D O N:
Printed for S. CHANDLER, *at the* Crofs-Keys *in the* Poultry.

Commissary of Supplies

Joseph McLellan of Falmouth, Maine.

Joseph McLellan was a Falmouth Neck man. He was the son of Brice and Jane McLellan and was born in Falmouth, in 1732. He married in September 1756, Mary McLellan a daughter of Hugh McLellan of Gorham, Maine,

Captain McLellan

He was a lieutenant in Capt. Joseph Pride's company in Colonel Joseph Prime's regiment at Falmouth, in 1780, and commanded a company from December 6, 1780, until May 1, 1781.

Capt. McLellan was first a mariner then a merchant. He was a selectman; county treasurer twenty-seven years, 1777-1803, and was on the committee to build the courthouse in 1787.

He was a prominent and respected citizen of Portland. His two sons, Hugh and Stephen McLellan were also Revolutionary soldiers and became prominent merchants of Portland.

He died July 5, 1820, aged eighty-eight years.

Remainder of the Command

Quarter Master General - Col. John Tyler.

Commissary of Ordinance - G. W. Speakman.

Deputy Quarter Master - Benjamin Furnass.

Dept. Com. of Ordnance - J. Robbins.

Brigade Majors:

Capt. Gowan Brown of Boston, and
Capt. William Todd.

Secretary - John Marston, Jr., of Methuen,
Massachusetts.

Quarter Master General - Col. John Tyler.

Commissary of Ordinance -- G. W. Speakman.

Deputy Quarter Master -- Benjamin Furnass.

Dept. Com. of Ordnance -- J. Robbins.

Fleet Captains - The majority of the other
captains of the fleet were privateers who, although
generally brave and experienced, focused on
gaining a profit in captured ships and cargos.
Successful privateers knew how to choose fights
they were particularly certain to win and profit
from.

Massachusets General Court Notified of British Occupation

June 24, 1779, Gen. Charles Cushing, of Pownalborough, Maine sent a letter to the Massachusetts General Court advising an immediate expedition to dislodge the British before they had time to entrench themselves. On June 25 the Court gave the Board of War directions to engage all state or national armed vessels that could be prepared to sail in six days.

The Board of War was also directed to charter or impress all private armed vessels available, with a promise to the owners of a fair compensation for all losses and damages they might sustain, and the wages of the men were to be the same as paid in the Continental service.

The Board was also to procure the necessary outfit and supplies, and the following were said to have been furnished:-- Nine tons of flour and bread, ten tons of salt beef, ten tons of rice, six hundred gallons of rum, six hundred gallons of molasses, five hundred stands of arms, fifty thousand rounds of musket cartridges with balls, two eighteen-pounders with two hundred rounds of ammunition, three nine-pounders with three hundred rounds of ammunition, four field-pieces, six barrels of gun powder, with a sufficient quantity of axes, spades, tents and utensils.

The fleet when ready would consist of nineteen armed vessels and twenty-four transports, all carrying three hundred and forty-four guns. The flagship was the *Warren*, a new thirty-two gun Continental frigate. The fleet would be under the command of Dudley Saltonstall of New Haven, CT, obstinacy outweighed his ability as a commander of a fleet.

Gen. Cushing of Lincoln County, and Gen. Samuel Thompson of Cumberland, were each ordered to detach six hundred men from the militia for two regiments, and Gen. John Frost was to detail three hundred of the York county militia to complete a sufficient number of men for the service to be performed. This would have made a total of fifteen hundred men, but in reality there would furnish less than one thousand or about the number of the enemy.

The resolve provided that such men as had been previously ordered to be raised in the above named counties, as a part of the state's quota of the Continental army, should be considered part of the said detachment and in case the expedition was carried into effect, the counties were to be exempted for nine mouths from raising men for the Continental service.

The militia for the expedition was collected with considerable difficulty. The reason given was that there was a misunderstanding of the meaning of the orders among the officers. Parson Smith, of Falmouth, records under date of June 30, 1779: *"People every where in this state spiritedly appearing in the present intended expedition to Penobscot, in pursuit of the British fleet and army there."*

Adjt. Gen. Jeremiah Hill testified at the investigation that *"the troops were collected with the greatest reluctance so that I recommended martial law. Some were taken and brought by force, some were frightened and joined voluntarily, and some skulked and kept themselves concealed. So upon the whole I collected by return four hundred and thirty-three rank and file."*

Adjt. Gen. Hill reported to Gen. Thompson the situation in Curnberland County and in reply, he said, *"If they will not go I will make the county too hot for them."* Brigade Major William Todd said that he marched to Casco Bay, July 14, with one hundred and thirty York County men, *"several of which were brought with force of arms."* He arrived at Falmouth the seventeenth, and found the transports waiting, and he testified at the investigation that there were   among the soldiers.

Col. Mitchell entered the service July 1, when he proceeded to collect and organize the men of his regiment. Their first parade together was July 8, and July 11, Parson Deane records that he "preached to the troops." Three days after Gen. Thompson wrote the following letter, probably to the Board of War:--

FALMOUTH, July 14, 1779.

Agreeably to your orders of the 26th of June last, I have detached out of my brigade 600 men, formed them into a regiment and appointed proper officers to command, viz: Col. Jona. Mitchell, Lt. Col. Nathan Jordan, Jacob Jordan first, and Nathaniel Larrabee second major. On the 6th inst. received orders from Genl Lovell to cause said troops to repair to Falmouth to be received by Major Hill who informed me he should be there the 8th, and on the 9th they would arrive at Falmouth and the greater part have been waiting ever since, except those who living near by had leave to return home for want of provisions.

SAM THOMPSON, Brig.

General Samuel Thompson then lived in Brunswick, and was the brigadier-general in command of the Cumberland county militia. He was a resolute, energetic and sincere patriot of the revolution, who for his early devotion to the cause of our independence, and his faithful public services, should be revered by the county of which he was then a citizen.

Ground Force Formed

On June 26, 1779, twelve days after the British landed at Castine, the General Court officially formed the ground force for the Penobscot Expedition. The Maine counties would provide the bulk of the force. The two closest counties, Cumberland and Lincoln Counties were to provide six hundred men each.

Three days later, York County received orders to add three hundred troops to the overall force. Expecting a short campaign, the General Court established two-month enlistments for the men raised for the expedition.

The final component of the army was a detachment of one hundred men from Lieutenant Colonel Paul Revere's Castle Island artillery unit. Revere's unit would employ the expedition's cannon consisting of two eighteen-pounders, three nine-pounders, and four four-pounders and one howitzer. The General Court directed the purchase of an extensive list of supplies and provisions authorized for the expedition.

American Revolutionary War: Private of Revere's Artillery Regiment 1779

The American Expedition is Formed

Assembling Expeditions

Prior to the Penobscot Expedition, Massachusetts showed a positive trend towards administrative proficiency of assembling seaborne expeditions to operate in Maine and Canada. While able to assemble necessary provisions and supplies, the militia still provided the manpower.

The colony came a long way from Sir William Phips ill fated attack on Quebec in 1690, where inadequate supplies proved his undoing. Fifty years later, when General William Pepperell's successful expedition to Louisbourg suffered similar although not as debilitating supply issues.

The fortunate interdiction of supplies by the Royal Navy boosted his siege as his supplies from Boston ran alarmingly low. Despite challenges in assembling the necessary provisions and equipment for the Penobscot Expedition, the only shortages noted by Lovell during the expedition were of hand grenades and mortars.

While shortages of food and equipment may have existed they did not hamper the American force. Nor were there are no existing records of the requisitioning or seizure of local supplies until the scattered westward retreat of the land and sea forces. This is a remarkable feat considering the competing demands of the ongoing war and necessity to provide supplies to regular forces serving in New York and Rhode Island.

To assemble its fleet, however, Massachusetts gambled its credit on the expedition's success. The colony had done so before and often lost, but the royal treasury became its safety net. Massachusetts may have assumed that the Continental Congress would do the same, and in hindsight they did. The impressive size of the fleet, despite its poor use, demonstrated not only the capability, but the commitment by the colony to build a successful expedition.

It is unlikely, however, that the magnitude of the disaster was s possibility to the General Court when it offered it. It was a calculated risk to get ships and crew cheaply to support the expedition. Besides the pay for the crew, the true financial incentive for ship owners was their anticipated shares in captured material and equipment.

While Massachusetts had developed the capability to build adequately supply expeditions, leadership experience of the militia constrained operations. While the colony built the expedition, it did little in the way of gathering intelligence, training the expeditions for specific tasks, and tactical planning; these areas remained up to the individual commander.

When Pepperell sailed from Boston, Governor Shirley had given him a detailed landing plan demanding precision timing. Realistic follow-on instructions remained vague and up to Pepperell. Even then, the friction of war rendered Shirley's plan worthless.

Commanders of provisional armies had to concern themselves more with the raising of the army than with the tactical issues that would face them on arrival. As a result, they often arrived on the scene with nothing more than a general landing scheme.

Commanders lacked the experience to think through entire campaigns. This is the case with Lovell when developing his plan for the Penobscot Expedition. While he developed a plan for the initial landing, he gave little thought as to how the follow-on objectives were going to be accomplished. It remained a constraint in both the militia's leader development and the provincial army system.

The Penobscot Expedition would be the last major expedition organized by Massachusetts. With the exception of small localized raids and the standing British garrison on the Bagaduce Peninsula, the Revolutionary War remained outside of Massachusetts' borders. Following the war, the national government assumed external security requirements for its states. Direct threats to Massachusetts from external forces eventually disappeared and the militia slowly absorbed into the national militia system even as universal military service declined. The debate of maintaining professional standing armies over militia forces remained an issue within America for years to come.

Preparing the Provincial Army

The inherent inefficiency of provincial armies frustrated those who tried to create and lead them.

While the colony often tried to select the most competent officers to lead the armies, they were often limited in the amount of experienced leaders to draw upon. Unfortunately, successful small unit tactical commanders rarely translated into successful commanders of large, complex armies.

General Pepperrell who led the successful Louisbourg Expedition in 1745 lacked any significant military expertise or training, but proved to be persistent and levelheaded, if not cautious. Other leaders such as Colonel John March, who led a small garrison in a successful defense against a French and Indian attack, found the challenges of besieging Port Royale at the head of a large army something he was ill trained and prepared for. After failing once and in the process of failing a second time, he broke down and handed off command.

General Lovell to Lead Land Forces

In selecting General Lovell, Massachusetts simply selected the best possible candidate for the position. As a member of the General Assembly, he was one the body's own and could be counted on to understand the importance of the operation.

Additionally, he had better military credentials than many. His performance in Rhode Island as a militia commander the previous year drew praise from commanders such as Nathaniel Greene and John Sullivan. Additionally he was a veteran of the French and Indian War and served early in the revolution, defending Boston. Few other leaders in the Massachusetts militia had such recent credentials.

General Wadsworth Chosen 2nd in Command of Land Forces

General Wadsworth was also well suited for his position. Only thirty-one years of age, he proved to be an energetic and able second in command for Lovell. For the most part, he led almost all major offensive actions during the expedition and more than any other leader, sought to salvage some sort of sanity during the chaotic destruction of the American fleet.

Lt. Colonel Revere to Command Artillery

As the only full-time officer, Lieutenant Colonel Revere for the most part fulfilled his duties as the commander of the expedition's artillery. He later ran afoul of General Wadsworth who ordered his arrest and accused him of dereliction of duty during the expedition's retreat.

Although he generally accomplished assigned tasks, he did so with little zeal.

While perhaps he was one of the few realists amongst the officers, records of the expedition's councils of war show him as an outspoken and early advocate of raising the siege and returning home without accomplishing the expedition's purpose. His actions during the expedition resulted in his removal from service.

The Maine militia that participated in the expedition is a somewhat unique blend of partially intact militia companies mixed into a provisional regiment. As a result, many of the men, volunteers or otherwise served under their normal company under their captain or one of an adjacent town.

Continental Marines 1775

Although some level of familiarity existed, it did not significantly enhance the effectiveness of the units. While the militia provided defensive functions for their towns, it did so in the form of small detachments of men performing garrison duties or watches.

Rarely if at all did units have the need to conduct company level training for defensive operations. At the regimental level, counties formed commands specifically for the operation, which is typical of a provincial army.

Most of the units formed for the first time a day or two prior to the arrival of the transports taking them the Penobscot Bay. Like more expeditions, the units assembled on the fly and attempted to sort things out as the different units and leaders attempted find out where they fit in the overall structure.

The other two land forces Lovell used during the expedition consisted of regular troops, or at least troops that served on a full time basis. Lieutenant Colonel Revere's detachment from the Castle Island artillery train consisted of soldiers with three-year enlistments that began well before the expedition. For the most part, these troops preformed their duties during the expedition strictly as specialists, primarily establishing artillery positions and employing the cannon.

The second force was the Continental and state marines that sailed with Saltonstall's fleet. The marines acted professionally and were well led down at the junior level. They formed the core of most of the expedition's major offenses. Marines conducted the first landings on Nautilus Island and served as a lead element in the landing on Dyce's Head.

The importance of the marines to Lovell is evident as it was a point of major contention between him and Saltonstall. It is unlikely that the final attack would have occurred at all if Lovell had not received one hundred marines from the fleet, even as they prepared to engage Mowat's ships themselves. Lovell's reliance on the marines speaks much to his assessment of his militia force.

Captain W. T. Welch Led the Marines

Marines under Captain Welch fought the sharpest battle of the expedition while climbing the rugged slope of the peninsula, suffering the heaviest casualties.

Capt. John Welch

Capt. Welsh and the Marines landed on Nautilus Island, routed the 20 British Marines on the Island, and installed a three-gun battery there.

Capt. Welsh Killed in Battle

At dawn on July 28, Capt. Welsh and his Marines, together with the Massachusetts militia, assaulted the mainland where the British were building their fort. The Americans, with the Marines on the right flank, landed under fire at the foot of a steep cliff and drove back the 300-man guard of British regulars. The Marines did the heaviest fighting. Capt. Welsh and eight Continental Marines were killed in the assault, and Marine Lt. William Hamilton was mortally wounded at the cliff's base.

Troop Recruitment

The task of recruiting the bulk of the army fell upon the commanders of the Maine regiments, General Charles Cushing of Lincoln County, General Samuel Thompson of Cumberland County, and General John Frost of York County. York County. General Frost, with the smallest quota, managed to raise only one hundred and thirty men of the three hundred ordered by the General Court.

Militia recruits

In Cumberland County, the Adjutant General Jeremiah Hill would report: *"the troops were collected with the greatest reluctance so I recommended martial law. Some were taken and brought by force, some were frightened and joined voluntarily, and some sulked and kept themselves concealed"*.

Parson Smith, of Falmouth, records under date of June 30, 1779: *"People every where in this state spiritedly appearing in the present intended expedition to Penobscot, in pursuit of the British fleet and army there."*

Adjt. Gen. Jeremiah Hill testified at the investigation *"the troops were collected with the greatest reluctance so that I recommended martial law. Some were taken and brought by force, some were frightened and joined voluntarily, and some skulked and kept themselves concealed. So upon the whole I collected by return four hundred and thirty-three rank and file."*

Militia marches to Casco Bay

Adjt. Gen. Hill reported to Gen. Thompson the situation in Cumberland County and in reply, he said, *"If they will not go I will make the county too hot for them."*

Brigade Major William Todd said that he marched to Casco Bay, July 14, with one hundred and thirty York County men, *" several of which were brought with force of arms."*

He arrived at Falmouth the seventeenth, and found the transports waiting, and he testified at the investigation that there were *"too many boys and aged" among the soldiers.*

Col. Mitchell entered the service July 1, when he proceeded to collect and organize the men of his regiment. Their first parade together was July 8,

Militia assemble

and July 11, Parson Deane records that he "preached to the troops." Three days after Gen. Thompson wrote the following letter, probably to the Board of War:--

FALMOUTH, July 14, 1779.

Agreeably to your orders of the 26th of June last, I have detached out of my brigade 600 men, formed them into a regiment and appointed proper officers to command, viz: Col. Jona. Mitchell, Lt. Col. Nathan Jordan, Jacob Jordan first, and Nathaniel Larrabee second major. On the 6th inst. received orders from Genl Lovell to cause said troops to repair to Falmouth to be received by Major Hill who informed me he should be there the 8th, and on the 9th they would arrive at Falmouth and the greater part have been waiting ever since, except those who living near by had leave to return home for want of provisions.

SAM THOMPSON, Brig.

In all, Cumberland County would muster four hundred and thirty-three soldiers of a quota of six hundred under the command of Colonel Jonathan Mitchell.

Fleet sets sail for Penobscot Bay

Lincoln County, which encompasses the Penobscot River, filled only half their quota. In total, eight hundred and seventy-two Maine soldiers mustered for duty.

After several delays caused by the many details of the expedition, the ships of the Penobscot Expedition departed Boston on 19 July, almost six weeks after McLean's force landed. The fleet sailed to Townsend, Maine to load the assembled militia onto the transports.

Arriving, at Townsend, General Lovell first learns the troops fall far short of their quotas, estimating the force consists of approximately a thousand men, not the sixteen hundred ordered by the court.

Following a review the next day, General Wadsworth disappointingly commented, *"At least one-fourth part appeared to me to be small boys and old men, unfit for service."*

Realizing that time was short, Lovell decided to continue with the forces available. After a delay caused by unfavorable winds, time which Lovell uses to drill the troops, the expedition departed on 24 July.

Meanwhile in Castine

While Massachusetts remained occupied with the details necessary for the Penobscot Expedition, General McLean's troops continued on fortifying the Bagaduce Peninsula. The original specifications called for a square fort, two hundred feet along the interior sides with bastions in the four corners.

Cannon at Ft George

Surrounding the fort was a ten foot high earthen wall in turn surrounded by a protective moat. Clearing trees and hauling in the necessary supplies delayed construction of the fort until 2 July when engineers began marking it out.

News of the Patriot expedition reached McLean on July 18th a day prior to its departure.

Captain Mowat was the first to respond to the threat, ordering his ships *"into the best situation to defend the harbour, annoy the Enemy and co-operate with the land forces."*

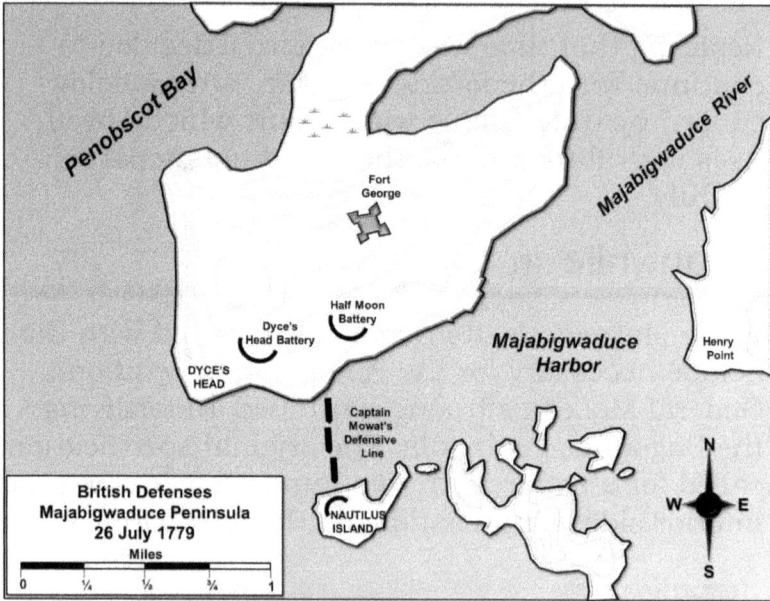

Map showing ships blocking entrance to Castine
Map created by Major Dale W. Burbank

Maximizing his three sloops-of-war, Mowat arrayed them across the narrowest point of the bay, anchoring his ships between the temporary fortifications on Dyce's Point and Nautilus Island. Through the use of spring lines, he ensured that all of his ships presented the broadside of his ships to any would-be attackers.

By the 21st of July, news arrived from a spy with the size and scope of the expedition. Detailed information on the expedition including the names of ships and their captain convinced McLean of its validity. McLean immediately transitioned from building the fort to making what he had thus far defensible. In a letter to General Clinton, he gives his bleak assessment of the fort on 21 July:

"Two of the Bastions were untouched, the other two with the Curtains were in general, from four to five feet and twelve thick, the ditch in many places not more than three in depth, And no Artillery mounted or Platforms laid; I had however some time before thrown up a Battery of four Twelve Pounders on a Height near the River for the protection of the Ships."

Realizing the gravity of the situation, McLean increased the force's efforts and Captain Mowat sent one hundred and eighty sailors ashore to assist in the preparations.

Earthen works at Ft George

Patriots Planning the Attack

While aboard the sloop *Sally*, General Lovell finalized his initial invasion plan. The plan looked to maximize the element of surprise, calling for an immediate large-scale landing of his forces upon arrival.

The plan, which did not account for any British warships in the harbor, was to land his forces in four main waves. Major Daniel Littlefield's detachment of York County militia would land first as the advance party and to protect the flanks of the first line as they landed.

The plan looked to maximize the element of surprise, calling for an immediate large-scale landing of his forces upon arrival.

The first line, commanded by Brigadier General Wadsworth consisted of the Colonel Mitchell's regiment of Cumberland County militia supported by one field piece. Once the first line disembarked, boats would return to pick up Colonel McCobb's Lincoln County militia regiment, also supported by one field piece.

Finally, Lieutenant Colonel Revere's artillery train would follow as soon as the conditions permitted. Given the minimal intelligence available to him, Lovell does not elaborate the plan of attack any further or actions taken once the troops land.

Expressing confidence in the abilities of his New England soldiers, the general assures that *"shuold there be an Opportunity [sic] he will have the utmost exertions of every Officer and Soldier not only to maintain, but to add new Lustre to the Fame of the Massachusetts Militia."*

On 24 July, a half-day's sail from the Bagaduce peninsula; Lovell's force still lacked intelligence on the British defenses. To rectify this deficiency, the commander of the *Tyrannicide* landed a small detachment of marines on the Fox Islands at the mouth of the Penobscot River.

Pretending to British sailors, the marines sought to gain some information on the British outpost. Somewhat confused by the ruse, the islanders reported what they knew about Mowat's ships and the incomplete state of McLean's fortifications.

Chief of the Penobscot Indians
"Bia Thunder" bv H.M. Burbank

The marines brought a few of the locals back to the ship for further questioning. Although British ships in the harbor were unexpected, the poor state of the fort was good news for Lovell. Earlier that day, Lovell received some unexpected allies against the British when forty-one Penobscot Indians approached the fleet in canoes.

After a brief council with Lovell on board, the Indians *"determined to proceed with us."* That night, orders went around to all of the ships to prepare for a general assault the next day.

The next day, the men aboard the transports prepared themselves for battle. The ships sailed up the Penobscot Bay arriving at the mouth of the Majawaduce Harbor at about noon. Captain Saltonstall found that Mowat's three sloops presented a formidable obstacle by arraying themselves broadside across the mouth of Bagaduce Harbor.

After a small council of war, Saltonstall formed up nine of his warships to attack Mowat's small fleet in three groups of three.

The shortcomings of the hastily assembled fleet quickly became evident. Ordered to join the attack, Captain Titus Salter of New Hampshire's twenty-four gun *Hampdon*, responded, *"I Should be Glad Sir If [sic] you would give me men a Nuf [sic] to man my ship."*

Meanwhile, Saltonstall had issues of his own. His inexperienced crew, over a hundred recently

Patriot's fire cannon on British at Castine

recruited for the expedition, showed their lack of gunnery skills. Calef, who witnessed naval gun battle from the shore wrote, "the fire of the enemy was random and irregular, and their manoevres [sic], as to backing and filling, bespoke confusion, particularly in the first division." The two-hour duel ended inconclusively with only minor damage to the rigging of the British ships.

Late that afternoon, despite Saltonstall's limited success against the British ships, Lovell began loading his troops into the small whaleboats for a landing. The rocky, brush-covered cliffs of Dyce's Head were not an ideal place for a landing, but did offer protection from the British ships. The thick woods would also offer cover for the landing troops.

At about seven o'clock, the first wave of whaleboats began rowing toward shore. High winds picked up and created large swells, making progress slow and difficult. Watching the ships struggle toward shore, Lovell became concerned that his first wave would be overrun before his boats could return with the second.

Patriot whaleboats struggle ashore

He decided to call off the assault just as the boats approached musket range. Struggling to turn the boats in the swells, the landing force lost one Indian to enemy fire.

The first day of the siege ended with little gain for the Americans, who began to identify some of the issues with the hastily assembled force.

The next day began with another naval attack that lasted for approximately two and a half hours. The positioning of the British ships in the harbor made it difficult for the American ships to close the distance between them. As a result, their fires were too far away to be effective.

American gunners focused largely on the British mooring points in an attempt to swing the ship away from their broadside positions. Consequently damage to the British ships was "chiefly to the rigging at the extreme ends of the ships."

Finding direct attack too difficult, Saltonstall planned to strike at the southern anchor point of the Royal Navy's line, Nautilus Island. At approximately five o'clock that evening, a force of two hundred continental and state marines under the command of Captain John Welch landed on the south side of Island.

The positions of the opposing forces

As a diversion, the militia's first line loaded their boats and feinted a landing Dyce's Head. Seeing the large invasion force of American marines supported by several warships, the twenty marine batteries on Nautilus Island quickly fled the island, leaving behind four small cannon and suffering one killed.

While the marines took the island without casualties, the feinting force lost one boat to a chance shot from a British battery, losing three men, including Major Daniel Littlefield, the commander of the York County contingent.

The Americans quickly added three heavier guns from Lieutenant Colonel Revere's artillery to the captured guns on Nautilus Island. As the Americans worked through the night to ready the guns, Captain Mowat quietly pulled back his ships out of the immediate range of the cannon and established a second defensive line.

The Americans spent the next day, 27 July, completing the battery on Nautilus Island and preparing for the main landings planned to occur at midnight. Unoptimistic that Saltonstall could dislodge the Royal Navy ships any time soon, he opted for a less than ideal landing at Dyce's Head.

Officers negotiate

Frustrated by the lack of action, thirty-two of the captains and lieutenants presented Saltonstall with a petition that morning.

The captains, mostly privateers, eager to finish the fight so they could move onto more profitable ventures, wrote;

" TUESDAY, A. M., 27 July, 1779.
" To the Honor 6 the Commodore & Commander-in-Chief of the Fleet now lying in this Harbor. A. Petition of the Lieutenants and Masters of the several armed vessels now under your Honours command.
Humbly sheweth. That we your Petitioners strongly impress d with the importance of the expedition, and earnestly desire to render our Country all the service in our power would Represent to your Honour that the most speedy Exertions should be used to accomplish the design we come upon.
We think Delays in the present case are extremely dangerous: as our Enemies are daily fortifying and strengthening themselves, & are stimulated so to do being in daily Expectation of a Reinforcement.
We don t mean to advise, or censure your past conduct, But intend only to express our desire of improving the present opportunity to go Immediately into the Harbour & attack the Enemy s ships, however we humbly submit our Sentiments to the better Judgment of those in Superior command. Therefore wait your orders whether in answer to our Petition or otherwise and as in duty bound will ever pray."

Signed by DAVID PORTER, 1st Lieut, of Ship " Putnam," and thirty others.

The American commanders hoped that the seizure of the Nautilus battery would give the American fleet a foothold into the harbor. From there, the fleet could attack the three ships, and support the landing. Instead, in a council of the war that afternoon, the American leadership decided on the opposite strategy.

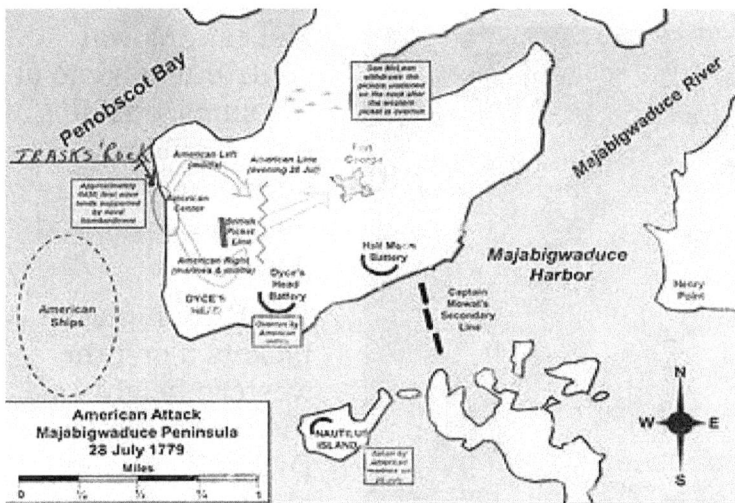

American's landing attack plan
by Major Dale W. Burbank

Land forces would first seize the peninsula and employ ground-based batteries against Mowat's ships. Dyce's Head on the west side of the peninsula remained the only viable option. To augment the landing forces, Saltonstall's ships contributed two hundred and twenty seven marines. Lieutenant Colonel Revere committed eighty men from his artillery train as well to support Lovell's eight hundred and fifty men from the militia.

With the addition of the marines to his landing force and the death of York County's commander, Lovell altered the first wave. Marines joined the first wave of attackers along with four hundred of the militia from the other counties.

For the landing, the fleet pressed every available boat into service. A small of flotilla of specially made flat-bottom boats, whaleboats, rowboats, and even one particularly flat-bottomed sloop loaded men for the assault.

While the Americans prepared for the assault, the McLean's men hastily strengthened their defenses. At the request of McLean, Mowat contributed some of his ships' offside guns to help strengthen the incomplete ground defenses.

Captain Archibald Campbell

The British placed pickets along the western heights of the peninsula on Dyce's Point and along the swampy neck to the north to protect against a mainland attack.

Manning the pickets on Dyce's Head stood Captain Archibald Campbell and his company of Hamiltons. An additional force of seventy men stood in nearby earthworks ready to support the picket line in the event of an attack.

The large granite boulder on the shore, now called "Trask's Rock," was named for a fifer-boy named Israel Trask, who took shelter behind it, playing his fife while his comrades made the ascent. It was said that he did not lose a note of the tune he was playing during the whole time. Capt. John Hinkley of Georgetown, of Col. McCobb's regiment, was killed while standing on this rock urging on the men.

The American Landing — 28 July 1779

At midnight, the landing force was still loading from ships to the boats waiting to take them to shore. Unable to sleep the night prior, many of the men conducting the assault were exhausted. These were the same militia and marines that conducted the attack of Nautilus Island and supporting feint the day before.

Finally at approximately four o'clock early the next morning, supporting ships began to shell suspected British positions on Dyce's Head. Meanwhile, the sailors began rowing the flotilla of

Americans attack the cliff

boats toward the shore. After about a half-hour of shelling, the American boats arrived along the rocky coast of the peninsula.

General Peleg Wadsworth led the first wave of attackers. Years later, he recalled the opening stage of the day's battle. *"The fire of the Enemy opened upon us from the top of the Bank or Clift, just as the boats reached the Shore. We step'd out & the Boats immediately sent back. There was now a*

107

stream of fire over our heads from the Fleet, & a shower of Musketry in our faces from the Top of the Clift. We soon found the Clift insurmountable even without Opponents. The party therefore, was divided into three parts, one sent to the right, another to the left till they should find the Clift practicable, & the Center keeping up their fire to amuse the Enemy."

Sent to the right was a contingent of mostly marines by Captain Welch mixed with a few of the militia. To the left was a detachment of militia, mostly from Colonel Mitchell's Cumberland County Regiment. In the center was a mixture of troops who attempted to draw the fire of the British and allow their comrades to the heights.

British soldiers hold their fire

For the defending British soldiers of the Hamilton Regiment, it was their first taste of combat. Inexplicably, the inexperienced company commander ordered his soldiers to hold their fire until the Americans landed, much to the benefit of the Americans.

Shaken by the American bombardment the British lines began to give way when the men saw a vastly superior number of enemy troops unloading the boats. Only one British officer, Lieutenant John Moore, continued to hold the line. Moore's detachment of twenty men anchored the picket's left flank. He soon found himself directly in the path of two hundred American marines and militia as they scrambled up the rocky heights.

Working along the shoreline, the marines soon found a steep trail leading off the right. Slowly they began climbing the heights, often using their hands to pull themselves up. Moore's Hamiltons poured musket fire into the hapless marines struggling up the heights, eventually killing nine, including their commander, Captain Welch.

American troops attack through woods

However, when the marines crested the first tier of the heights, it was Moore who came under heavy fire. Moore later wrote, *"I confess that at the first fire they gave us, which was within thirty yards, I was a good deal startled, but I think this went gradually off afterwards."*

As he fought off the marine attack, Moore found himself between the American militia to his right and the marines circling to the left. Even though seven of his men were dead and several more lay wounded, he continued to fight.

Meanwhile, Moore's commander who fled with the bulk of his troops reported to General McLean. Hearing sounds of continued fighting, a skeptical McLean sent a fresh company back to the picket line to attempt to salvage the situation. Recognizing the precarious position Moore's men were in, the relief commander directed him to withdraw or face complete envelopment by the Americans.

After Moore's detachment retired back to Fort George, the marines found themselves in command of the high ground of Dyce's Head. Only twenty

Americans Charge artillery position

minutes had passed since they first hit the shore. After pausing briefly to regroup the men, they moved out on a broad front over the high ground and into the thick woods on the other side. Cresting Dyce's Head, the marines noticed a small artillery battery at the bottom of the point along the shoreline, currently dueling with American ships and the battery on Nautilus Island.

Charging down the hill, they quickly routed the British artillerymen who quickly retreated to the safety of Mowat's ships anchored in the harbor.

110

Calef wrote that *"an attempt was made to demolish the guns, but the enemy would not suffer it."*

The Americans captured the entire battery's compliment of three guns intact. Soon afterwards, the Americans had secured the entire western portion of the Bagaduce Peninsula. Even the picket stationed along the neck of the peninsula withdrew for fear that the Americans would cut them off.

As the Americans continued to sweep across the central plateau, McLean grimly organized his defenses. He knew his half-finished defenses would not be sufficient to halt a determined American attack.

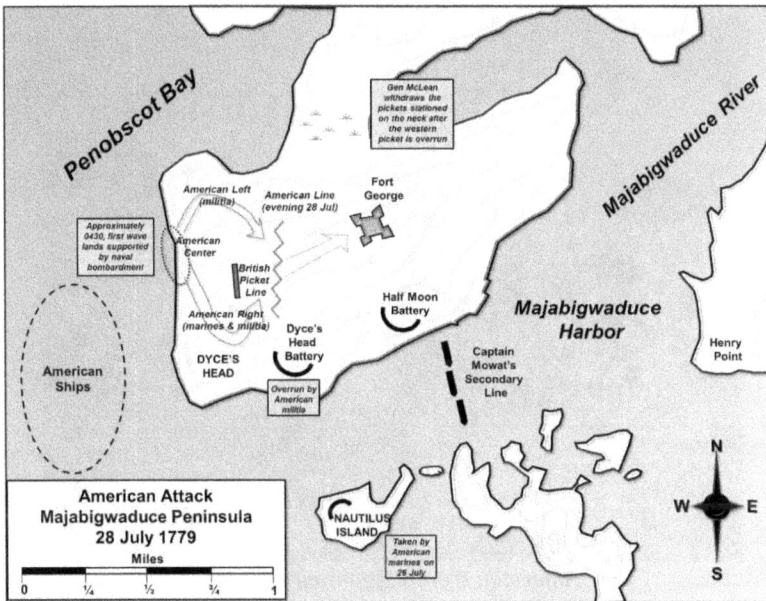

Map summarizing the attack by Major Dale W. Burbank

He appointed Lieutenant Moore the commander of a fifty- man reserve force with orders that *"should the enemy rush forward, as soon as they got into the ditch of the fort, he should sally out and attack them on the flank with charged bayonets."*

In reality, he thought the battle had already been lost. After the battle, McLean would remark, *"I was in no situation to defend myself, I only meant to give them one or two guns, so as not to be called a coward, and then have struck my colors, which I stood by for some time to do, as I did not wish to throw away the lives of my men for nothing."*

The Americans, however, were far less aggressive in their attack than McLean gave them credit for. As the men reached the edge of the cleared space around the fort, they stopped. With no orders or leaders encouraging them to go farther, they instead stopped their forward attack.

American troops stop their attack

General Lovell seemed pleased with the progress, reporting back to the Massachusetts General Court that *"we are within 100 Rod [550 yards] of the Enemy's main fort on a Commanding piece of Ground, & hope soon to have the Satisfaction of informing you of the Capturing of the whole Army."*

American casualties were much lighter than expected, with sixteen killed and twenty-one wounded. Perhaps surprised by their success, the Americans failed to press home the attack. Instead, they returned the initiative back to the British and called forward the entrenching equipment and Revere's cannon.

At about ten o'clock, Captain Saltonstall decided it was time to engage the Royal Navy ships in the harbor. Using his ship, the *Warren* and three other ships, he engaged the British ships from long distance from the mouth of the harbor.

Supported by the Nautilus Island Battery, Saltonstall kept up *"abrisk fire"* for a half an hour.

USS Warren fires upon Castine

However, the British gunners got the best of his ship, crippling it and forcing Saltonstall to retire after a brief thirty-minute exchange. The *Warren's* mainmast was shot through in two places in addition to other significant damage.

Saltonstall's inexperienced crew nearly ran the fleet's flagship aground before they managed to anchor the ship safely. Extensive repairs rendered the ship incapable for the next two days.

The Royal Navy ships under Captain Mowat accounted for themselves well, but the foothold on the peninsula combined with the continued fire from Nautilus Island forced the ships to again move further back into the bay.

A Siege Instead of Victory

Apparently unaware of how close he was to victory, General Lovell was happy with the day's attack, noting in his journal that *"scuh a landing has not been made since Wolfe."*

The American landing achieved surprise and threw the British garrison off balance. He also isolated McLean's force from the mainland by seizing the peninsula's neck and the entire western half of the peninsula was now in American hands.

Americans begin preparations for siege of Castine

Although the Royal Navy kept the American fleet at bay, the Americans contained them within the Bagaduce Harbor. Cut off, the only hope for the British was to hold until relief arrived. Success for the Americans meant using their superiority of numbers, before it was too late. Instead, the Americans settled into a siege that they could hardly afford.

With great industry, the Americans began work on fortifications from which they could lay siege to the British fort. American militia soon found themselves busy cutting roads, clearing fields of fire, and digging entrenchments.

Seamen ferried cannon from the ships to shore and work details manhandled the cannon up the hill and into positions established by Revere's men. In two days, by 30 July, the Americans had two eighteen pound and one twelve pound cannon along with a five inch mortar in action against the British.

The Americans built series of zigzag trenches to protect their lines and to inch the guns closer to the fort. The recently captured battery on Dyce's Head joined in with the one on Nautilus Island in an almost continual bombardment of the British ships in the harbor.

Americans fire on the British ships in Castine Harbor

Successfully landing their force, the Americans needed to figure what to do next. From Lovell's perspective, he needed the support of Saltonstall's fleet to storm Fort George. For Saltonstall, his attacks on the British ships had proved inconclusive.

Concerned by the confining geography and difficult currents of the bay, he was reluctant to commit his ships in an all-out attack on Mowat's three ships. For Saltonstall, it was clear that Lovell's men had to silence the land-based gun batteries before he could safely attack the British ships.

A council of war aboard the *Warren* comprised of the fleet's captains on the 29th echoed this sentiment. The reduction of the remaining battery at the entrance to the harbor, known the *"half-moon battery,"* became a precondition to naval attack. Meanwhile, Lovell and his commanders had other distracters.

With the militia off the boats, discipline issues began to ripple across the American camp. On 30 July, General Lovell issued orders designed to curb what he termed as *"alarming behavior amongst the men"*.

The first, designed to curb an increasing desertion rate, prohibited non-commissioned officers and soldiers to *"presume to be more than twenty Rods [110 yards] absent from his Lines."*

His second order sought to enforce greater discipline, directed that the men were *"strictly forbid to fire their Guns in such a loose unsoldierlike [sic] manner as have been practised [sic] of late."* To add to the expedition's woes, news arrived of impending British reinforcements.

British Reinforcements on the Way

The *Diligent*, dispatched to patrol the coast, had captured a British schooner carrying messages for General McLean at Castine. Although the dispatches were destroyed, several prisoners from the ship provided credible evidence that a British fleet was assembling to sail to McLean's relief.

The "Diligent" captures British Schooner

Finally, after several days of inactivity, Lovell formed a plan of action. With returns showing that he had 873 men militia available for action, he worried that his force was too small to take fort. He looked locally for new reinforcements to grow his army. To do so, Lovell issued a proclamation similar to McLean's a month earlier. In his proclamation, he gave local residents forty-eight hours to join his camp with *"such arms and accoutrements as they now possess."* While evidence suggests he did entice some men to join his ranks, they served only to offset his losses due to desertion, illness, and combat related casualties.

For the second part of his strategy, he planned an attack on the half-moon battery, hoping that in taking it, he could finally get the support of Saltonstall's fleet. His plan called for a nighttime attack with a force of three hundred soldiers, marines, sailors, and Indians under the command of Brigadier General Wadsworth.

In the early evening hours, Wadsworth led his force to a hiding site near the battery. There, he waited until approximately two in the morning, when he suspected the redoubt's defenders were least alert, and rushed its fortifications.

Indian scalps British dead

Although the fifty British marines put up a good fight, the American quickly overwhelmed them. Half of the defenders managed to reach the safety of the main fort. Of the other half, five lay dead and fifteen were either wounded or surrendered. American losses were five killed and another twelve wounded.

As the Americans consolidated their gains, the Indians that joined the party reportedly stripped and scalped the British dead.

Understanding its importance, British General McLean prepared a counter-attack. At about five o'clock, fifty men from "The Hamiltons," charged the American's battery with bayonets fixed. Unprepared for a determined counterattack, the charge drove the entire three hundred man American force from the battery. Barely managing to destroy some supplies and pausing long enough to knock the cannon from the mounts, the Americans retreated back to their fortifications.

Wadsworth's rout, combined with a heavy rain that began at dawn, dealt a significant blow to morale.

Lovell needed to find another way to get Saltonstall's fleet into the fight. For the first time, Lovell admitted to the General Court that *"the Enemy cannot be attacked by Storm with any probably of success-their works being exceedingly strong and our Troops (tho brave) are yet undisciplined."* The next day he requested *"a few [hundred] regular disciplined troops and Five Hundred hand Grenades from the Massachusetts General Court to allow him to take the fort by force."*

British counter attack

General Lovell Refused to Make Contingency Plans

Despite the growing realization that he might not succeed, Lovell refused to make any contingency plans in the event British relief arrived. General Wadsworth, who was earning a solid reputation amongst the men by leading the last two major assaults, entreated Lovell to make plans in the event the expedition failed.

It was urged upon Genl Lovell *"to erect some Place of resort up the River at the Narrow, in Case of Retreat so that the Troops might have a place of resort in case of necessity & also to have some place of Opposition to the Enemy should He push us thus far-but the Genl would hear nothing of the kind: alledging [sic] that it would dishearten our Army & shew [sic] them that we did not expect to succeed."*

While morale amongst the men was certainly an issue, in hindsight, the lack of planning proved to be disastrous.

Unable to leverage the naval fleet to provide a decisive edge against the fort, General Lovell next chose a strategy of slow attrition, slowly nibbling at the perimeter of the British forces.

Men dragged guns across the peninsula's neck and constructed a battery of three guns at the Wescott house across the muddy flats that separated the mainland from the peninsula. Built directly to the north of Mowat's ships, the battery enabled the Americans to put effective fires on the British ships.

The new Wescott Battery, put in action on 4 August, complemented the existing battery on Nautilus Island firing from the west. Seeking to put the British ships in an even less tenantable position, General Wadsworth led a detachment across the Bagaduce River to establish another battery at Hainey's Point.

From this vantage point, the Americans could place the British in a three-way crossfire. Exposed to the British ships, work on the redoubt proceeded slowly. On 9 August, the HMS *Nautilus* discovered one of the American boats ferrying supplies across the river to the new battery. Realizing the gravity of the situation, Mowat dispatched a force of fifty sailors and soldiers to attack. Too isolated from other American support, Wadsworth abandoned the construction of the battery.

Although he achieved limited success at eroding the British perimeter, Lovell was running out of time. In addition to the threat of British reinforcements, attrition through constant shelling, illness, and desertion was taking a toll on his army.

The numbers of men declined from 847 men fit for duty on 31 July down to 715 on 7 August available for duty.

Frustrated by the lack of progress, he wrote Commodore Saltonstall on 5 August, *"I have proceeded as far as I can on the present plan and find it inafectual [sic] for the purpose of dislodging or destroying the shiping I must therefore request an ansure [sic] from you whether you will venter [sic] your shipping up the river in order to demolish them or not that I may conduct my selfe accordingly."*

His answer came the next day during a council of war aboard the *Warren*. The ship captains agreed that an assault of the fort was a necessary precondition to any action against the British ships.

No Way the Fort Could be Taken by Storm

On the 7th of August, Lovell met with both his own officers and those of the naval contingent. The council unanimously decided that the enemy fort could not be taken by storm. Still trying to salvage something from the expedition, Lovell proposed a plan to take the eastern part of the peninsula. In doing so, he could isolate the British fort and from the supporting ships.

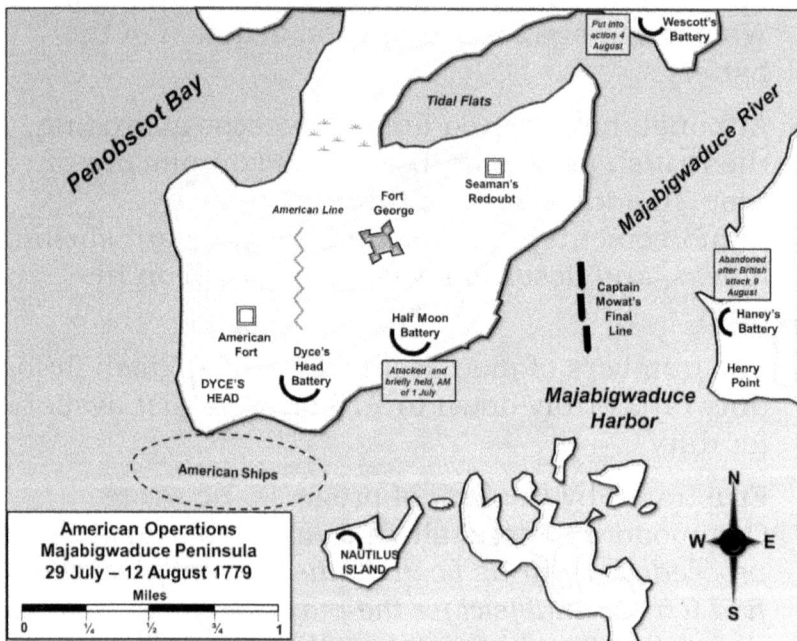

American positions after initial attack – by Major Henry W. Burbank

Since August sixth, seamen from Mowat's ships went ashore to construct a small redoubt (a temporary or supplementary fortification, typically square or polygonal) on the east side of the peninsula. Designed primarily to protect the rear of Fort George, the redoubt also provided a refuge for the sailors and marines if their ships were lost.

Approximately three hundred sailors and marines alternatively pulled ship and shore duty based on the perceived American threat. By isolating the land and navy components, Lovell hoped Saltonstall could destroy Mowat's fleet.

Cut off from external support, Lovell hoped the fort's defenders would have to surrender. The plan did not convince all of his officers. For the first time, members of the council, most notably Lieutenant Colonel Revere talked openly of abandoning the siege. It has been reported that Revere had actually voted to go home!

When the council came to an inconclusive end, the expedition was no closer to determining a plan of action.

After the council, hoping to demonstrate progress, Lovell ordered his men to conduct a feint to the south, hoping to draw out and then ambush British defenders from the fort. Hiding in the woods nearby, lay the one hundred militia soldiers of Lovell's ambush force. Initially, it appeared the ruse would work. Lovell noted, *"the bait took, they soon sallied [sic] with 80 Men & rush'd down to cut off our parties."*

Instead, a determined attack from the British soldiers routed both the baiting and the ambush forces. The Americans ran back into friendly lines without causing any significant harm to the British soldiers. It was not the demonstration of American arms that Lovell had hoped for and served to sap the American's morale.

The next day, however, Lovell received an unexpected ally in Saltonstall's fleet. Captain Hoysteed Hacker of the Continental Sloop *Province* finally broke ranks with his superior. While he could not control the opinions of the privateer captains, Saltsonstall had always counted on the unquestioned support of his subordinate captains from the navy.

Unable to stand the inaction any longer, Hacker spoke out. In a letter to both Saltonstall and Lovell, he presented a plan that complimented Lovell's proposal for a coordinated land and sea attack. He suggested that Lovell's men augmented with one hundred marines attack the rear of the peninsula as three of the American ships attacked the British ships.

Five other warships conducting a diversionary attack upon the main fort would cover these three ships. Hacker's letter provided the spark needed to convince Saltonstall to commit to an attack. Two days later, a council of war unanimously approved the plan. Now it was up to Lovell to prepare his troops to do their part.

Lovell ordered his regimental commanders to call for a six hundred man volunteer force from their units. While he gave each of the regiment a quota, only one managed meet its quota of 200 men. The York regiment could not meet its quota because it had sent too many men chasing after twenty of its deserters. The other regiment faced similar issues.

Hoping to increase the confidence of the men, he again tried a demonstration designed to draw the defenders from the fort. Two hundred men marched out the now abandoned half-moon battery. His soldiers made a show of refortifying the battery, even sending out work details in hopes of drawing out some of the defenders.

Unbeknownst to Lovell, one hundred of the British did come out, but hid in the nearby cornfields, preparing for a nighttime attack. At dusk, Lovell convinced he was unsuccessful in his endeavor, recalled his men. As the Americans returned to their defenses, the British attacked, routing them with minimal casualties on either side.

As Lovell struggled to find a successful strategy on the Bagaduce Peninsula, the General Court was busy trying to find him reinforcements.

General Washington, at his headquarters at West Point, received intelligence that a British relief force sailed for the Penobscot. Concerned, he relayed the information to Massachusetts on 3 August.

Meanwhile, Major General Horatio Gates dispatched a regiment of four hundred seasoned men under the command of Colonel Henry Jackson to the Penobscot. Although promising, these troops needed to move from their current location in Rhode Island to the Penobscot.

"as soon as you receive this you must take the Most Effectual Measures for the Capture or destruction of the Enemies Ships & with the greatest dispatch the nature & Situation of things will Admit of."

The Massachusetts Court warned Lovell not to delay in hopes that Jackson's men would reach him before the British received their reinforcements, warning that *"your situation is very critical. Something must be hazarded and Speedily too."* The Navy Board was more direct in their instructions to Saltonstall. In a dispatch on 12 August, the board directed him *"that as soon as you receive this you must take the Most Effectual Measures for the Capture or destruction of the Enemies Ships & with the greatest dispatch the nature & Situation of things will Admit of."* It became apparent to all parties that time was running short on the Penobscot Expedition.

Lovell's Final Attack and the Destruction of the American Fleet

On the 13 August, the sixteenth day after the initial landing, the Americans prepared for their coordinated attack. Lovell would attack first, sweeping around to the east side of the peninsula and cutting off Mowat's ships from supporting the fort. Once he was in position, Saltonstall's ships would attack.

That morning, Lovell held a final council of war with his field officers. Even on the day of attack, the council remained divided. When the meeting concluded, the officers brought the issue to a vote.

Nine voted to evacuate completely, while thirteen voted to *"tarry on the peninsula"*. Just after noon, Lovell personally led the attack recognizing that the attack may decide the entire campaign. Lovell's two hundred man force ran a gauntlet of cannon fire from the fort, but emerged on the other side of the peninsula without opposition.

The Americans had the British trapped; all that was left was for Saltonstall's fleet to finish the job and then . . .

The Americans now held all but the center of the peninsula and its harbor. With his portion of the attack completed, an ecstatic Lovell sent a messenger to Saltonstall to begin his attack. It was almost five o'clock when Saltonstall began his final and hopefully decisive attack.

Just as the ships made way, the American and British combatants spotted strange sails coming up the bay from the south. The American fleet halted its attack, trying to determine if the ships were American or British reinforcements. Soon

126

became painfully clear that the long awaited British relief force arrived.

Attack called off and troops retreated to their ships as the British Relief Force approached

British Relief Force arrives as attack on Castine is underway

At sunset, as dejected General Lovell returned back to the American lines. At midnight, Lovell ordered his commanders to prepare their troops and equipment to depart. By three o'clock the next morning, Colonel McCobb's men began loading their equipment onto boats. Two hours later, the men loaded the boats taking them to the transports.

Once loaded, the transports sailed up the Penobscot Bay and away from the newly arrived British fleet. At about four that morning, McLean noticed the American positions were unusually quiet and ordered a reconnaissance of their positions and found them thoroughly abandoned. Americans even evacuated all of the cannon with the exception of the guns on Nautilus Island, which they spiked and left in place.

With the withdrawal from the peninsula complete and all men and equipment safely aboard, the transports slowly made their way up the bay towards the mouth of the Penobscot River. There, although blocked from the open sea, the men could unload and establish an effective defense from the British attackers.

Saltonstall's warships arrayed in an arc appearing to establish a defensive line designed to protect the transports as they sailed up the bay. The morning was calm with little wind, so all the ships sat becalmed for several hours.

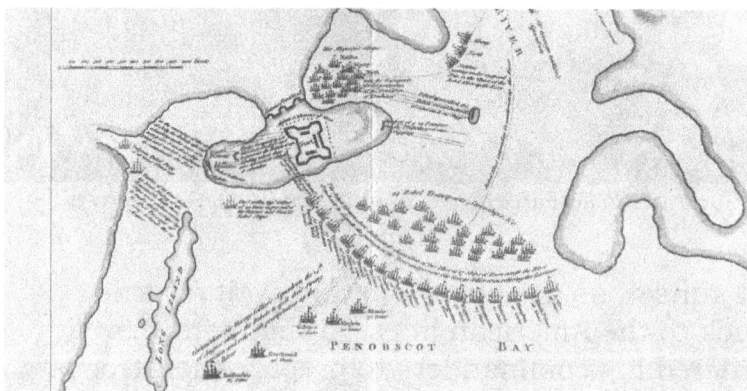

Americans establish defensive lines as British approach

Eventually, the British fleet under the command of Sir George Collier began slowly inching his way towards the seventeen American ships with his six larger warships. The Americans waited, presenting their broadsides to the British who sailed in two echelons of three.

Saltonstall orders his fleet to *Run Away*

Although he outnumbered the British ships, Saltonstall chose not to hold the line. As the British fired their first salvos at the American fleet, Saltonstall, aboard the *Warren*, ordered the dispersal of the fleet. It would be his last order to the fleet.

From this point, every ship fended for themselves. The fastest of the American ships, the *Hunter*, attempted to escape past the British ships and into the ocean. The British flagship, the *Raisonable*, intercepted and captured her. The remainder of the ships sailed up the bay, many passing up the slower transports.

Only one ship, New Hampshire's *Hampden* put up a fight. Inexplicably sailing poorly, the British warships quickly caught her. The *Hampden* fired upon the British ships for thirty minutes doing little damage until crippled by point blank-fire from a British warship. With several men wounded, the captain finally surrendered.

New Hampshire's Hampden does battle with the British

Further up the bay, the remains of the American fleet were in full retreat towards the comparative safety of the Penobscot River. A light breeze coupled with an outgoing tide made progress difficult, if not impossible for the slower transports. Unable to continue progress up the bay and in some cases moving backwards on an outgoing tide, captains began beaching their ships, setting them on fire before running into the woods.

General Wadsworth attempted to salvage the situation and establish a defensive position at the mouth of the Penobscot River. He moved up and down the bay, gathering men and sending them north to the Penobscot River. He attempted to salvage key equipment and secure cannon from Revere's ship to build his defense.

Despite his best efforts, the Penobscot Expedition simply just fell apart. A sense of panic engulfed the already poor morale of the force. By the morning of the sixteenth, crews set the last of the ships on fire and headed east into the woods.

Nineteen year-old Thomas Philbrook, one of the marines of the expedition, summed up the final days of the expedition.

Our retreat was as badly managed as the whole expedition had been. Here we were, landed in a wilderness, under no command; those belonging to the ships, unacquainted with the woods, and only knew that a west course would carry us across to Kennebec; whereas, there were hundreds of militia that were old hunters, and knew the country. Some of these ought to have been detained as pilots, and we might have got through in three days; but we had no one to direct; so everyone shifted for himself.

Aftermath

The American losses were heavy in terms of equipment but hard to measure in terms of personnel. The British fleet ensured the capture or destruction all seventeen warships and twenty transports present when it arrived.

For Massachusetts, who underwrote the losses of the private ships, financial liabilities exceeded 1,041,760 pounds. After petitioning the Continental Congress for several years, it finally agreed to reimburse the wartime losses.

By 1781, however, Massachusetts' credit became so poorly rated it had to pay cash when purchasing items.

American casualty numbers vary widely due to the disorganized retreat and the increasing desertion rates during the later stage of the expedition. Based on primary sources, actual battle losses numbered about one hundred men.

Some historians estimated that the destruction of the fleet and the subsequent retreat resulted in another three hundred American casualties, many of which lost their way in the wilderness and died of starvation.

Colonel Jackson's regiment arrived in Falmouth as the details of the breakup of the Penobscot Expedition began arriving. For Jackson's men, their mission became a much different task of defending the Maine settlements from further British expansion. By October, it was apparent that an attack was no longer imminent and they departed, handing the security mission to General Wadsworth and the Massachusetts militia.

The disastrous end of Penobscot Expedition sparked the General Court to convene a committee to investigate the causes of failure. The committee questioned or received testimony from most of the key participants except for Captain Saltonstall, who as a Continental Navy officer remained outside of the board's jurisdiction.

During the War of 1812, British forces destroyed the naval archives that held the proceedings of Saltonstall's court martial, and no other copies of his version of events survives. The Massachusetts committee released its findings less than two months later on 7 October 1779.

Three of the findings, including the principal cause of failure, placed the blame directly on Saltonstall. His "want of proper spirit and energy‖ was the primary cause of failure, further aggravated by his discouraging any offensive operations by the American fleet and for —not exerting himself at the time of the retreat, in opposing the enemy's foremost ships."

There are certainly elements of truth to the findings. Notwithstanding the tactical problems he faced with the Bagaduce Harbor, his failure to plan for and organize a defense against the British relief fleet is inexcusable. For Massachusetts, blaming Saltonstall for the failure strengthened their claim for reimbursement from the Continental Congress since it was their officer that led the expedition to defeat.

The committee found General Lovell acted with *"proper courage and spirit‖ and with the support of the Commodore and the appropriate number of men ordered for service would have been successful."* The board commended General Wadsworth, perhaps the competent of the senior leadership his *"gerat activity, courage, coolness and prudence."*

The committee only vaguely admonished the leaders of the Maine militia for failing to raise the full complement of militia ordered by the General Court, assuming that many troops were available to muster.

Root Causes of Failure

In terms of pure numbers, Lovell's army matched fairly evenly with McLeans in both numbers of men and in artillery.

From the sea, the Americans had overwhelming naval superiority. Yet the leaders of the expedition failed to agree on a method to leverage their advantages against the British.

The element of surprise, compounded by the psychological effect of the large American fleet, was almost enough to win the day on 28 July 1779. Poor leadership and cooperation continued to keep victory out of reach. While much of the blame for the disaster falls on its leaders, there are deeper causes that lay within the colony's provincial army system.

While the Penobscot Expedition shared many characteristics with Massachusetts's provincial armies, a total call-up of the militia created its own challenges. Members of the expedition more closely resembled the general militia at large and not the normal recruiting pool for provincial armies.

While the average age of men in provincial armies ranged from the teens to mid thirties, the militia had men aged sixteen to sixty-four. Thus it is likely that General Wadsworth's comment that a quarter of the men mustered for the expedition were boys and old men was true.

Despite the employment of relatively intact militia companies, units still lacked the same training and leaderships skills of the provincial armies.

133

Penobscot Expedition Warships

American
Active – 16 Guns
Black Prince – 18 Guns
Hampden – 20 Guns
Surrendered after "finding herself closely beset"
Putnam – 22 Guns
Monmouth – 24 Guns
Warren – 32 Guns
Vengeance – 24 Guns
Sally – 22 Guns
Hector – 20 Guns
Sey Rocket – 16 Guns
Hunter – 18 Guns
Defence – 16 Guns
Hazard – 16 Guns
Diligence – 16 Guns
Tyrannicide – 14 Guns
Providence – 14 Guns
Spring Bird – 12 Guns
Burnt at the mouth of the river

British
Nautilus - Sloop
Albany - Sloop
North - Sloop
Rationale – 64 Guns
Greyhound – 20 Guns
Galatea – 20 Guns
Virginia – 20 Guns
Blonde – 32 Guns
Camilla *Cap* Collins – 20 Guns

British

Journals, Proclamations and Letters

Written by those who where there

British Officer's Summary of the Expedition

"Upon the 10th of June the colonels M'Lean and Campbell arrived from Halifax with 450 men of the 74th and 200 82nd of the regiments under convoy of the Albany, Natilus, and North sloops of War in order to establish a port upon the Penobscot River; till the 25th of July they were employ'd in clearing the Ground and Constructing a Fort which they had not half compleated when Commodore Saltonstall's fleet of 17 ships of War and 24 Transports with 2500 Land Troops on board under the command of General Lover arrived from Boston, from the 26th of July they cannonaded our Fort and Shipping which they expected to make themselves masters of by a General Attack upon the 14th of August but upon Sir George Collier's appearance that day with a squadron of 5 Men of War from New York; they abandoned their Works and retired on board their Fleet, they made a Show of disputing the passage were drove up the River and totally destroyed.

During the siege His Majestry's Sloops of War the North, Natilus, and Albany, lost only 15 Men killed and Wounded, and the 74th and the 82nd Regiments 70 Men killed and Wounded.

The Losses the Rebels sustained is unknown.

PROCLAMATION – June 15, 1779

By Brigadier-General FRANCIS MCLEAN and ANDREW BARKLEY, Esq., Commanding detachments of his Majesty s Land and Naval Forces in the River Penobscot.

WHEREAS it is well known that there are in the several Coloniest in North America, now in open rebellion, many persons who still retain a sense of their duty, and who are only deterred from an open profession of it by the fear of becoming objects of the cruel treatment which they have seen exercised on others, by persons who having plunged their country into the horrors and distresses it now labours under, industriously seize every opportunity of gratifying their avaritious and wicked dispositions by the wanton oppression of individuals:

And whereas it hath been represented that the greater part of the inhabitants on the river Penobscot, and the several islands therein, are well affected to his Majesty s person and the an cient constitution under which they formerly flourished, and from the restoration of which they can alone expect relief from the distressed situation they are now in:

Their Excellencies the Commanders in Chief of his Majesty's naval and land forces in North-America, taking the good dispositions of the inhabitants above mentioned (as represented to them) into their consideration, and desirous of encouraging and protecting the persons professing them, and securing them from any molestation on that account, have ordered here the forces under our respective commands for that purpose: We therefore, in obedience to their directions, hereby invite and urgently request the

137

inhabitants on the river Penobscot and the islands therein in general, to be the first to return to that state of good order and government to which the whole must in the end submit, and openly to profess that loyalty and allegiance from which they have been led to swerve by arguments and apprehensions, of the falsehood of which they must have been long ago sensible, as well as of the views of those who first promoted them. We also call on all those whose principles have never been shaken, to embrace the present opportunity of manifesting them without dread or apprehensions, as we hereby assure them of every protection in the power of the forces under our respective commands to bestow. And, to quiet the apprehensions of any persons who might be deterred from embracing this opportunity by the dread of being punished for any former acts of rebellion which they may have been led to commit, we hereby declare that we will extend our protection, and give every encouragement, to all persons of what ever denomination who shall, within eight days from the date hereof, take the oaths of allegiance and fidelity to his Majesty, before such persons as we shall appoint, either at the head quarters of his Majesty s troops at Majabigwaduce Neck, or at Fort Pownal* ; which oaths of allegiance and fidelity we require all persons whatever to come and take within the required time, and not, by neglecting to give such testimony of their loyalty, give room to look upon them as desirous of continuing in an obstinate and unavailing rebellion, and subject themselves to the treatment such conduct will deserve.

To all persons who by returning to their allegiance shall merit it, we not only promise protection and encouragement, with the relief that shall be in our power to alleviate their present distresses, but we also declare that we will employ the forces under our command to punish all persons whatever who shall attempt in any manner to molest them, either in person or property, on account of their loyalty or conduct toward us; and if forced by their behaviour to punish any men or set of men, on the above-mentioned account, we declare that we will do it in such an exemplary manner as we hope will deter others from obliging us to have recourse to such severe means in future.

And whereas the inhabitants to whom this proclamation is addressed, as well as those in general settled in that part of the country called the Province of Maine, have settled themselves on lands, and cultivated them, without any grant or title by which their possession can be secured to them or their posterity ; we therefore declare that we have full power to promise, and we do hereby promise, that no person whatever who shall take the oaths of allegiance as above required, and give such other testimony of their attachment to the constitution as we, or other officers commanding his Majesty s forces may require, shall be disturbed in their possessions; but that whenever civil government takes place, they shall receive gratuitous grants from his Majesty (who alone has the power of giving them) of all lands they may have actually cultivated and improved.

And whereas the leaders of the present rebellion, in pursuit of the views which first instigated them to foment it, and probably to blind the people with regard to the cause of the severe distress under which they now labour, have industriously propagated a notion that the officers of his Majesty s sea and land forces willingly add to their sufferings: We, therefore, to remove such prejudices and as far as in us lies to alleviate the misery of the inhabitants of the villages and islands along the coast of New England, hereby declare that such of them as behave themselves in a peaceable, orderly manner, shall have full liberty to fish in their ordinary coast fishing craft without any molestation on our part; on the contrary, they shall be protected in it by all vessels and parties under our command.

Given on board his Majesty s ship "Blonde", in Majabigwaduce River, The 15th of June, 1779.

FRANCIS McLEAN,

ANDREW BARKLEY.

A British Journal

The Journal

June, 17, 1779, Brigadier-General Francis McLean landed at Majabidwaduce (Penobscot), with about 700 of his Majesty s forces, composed of detachments from the 74th* and 82d **regiments, to take post in the eastern country of New England. The time from this day to the 17th of July was taken up in clearing a spot to erect a fort and building the same, and a battery near the shore, with store-houses, etc.

July 18. Intelligence was received that a fleet and army were preparing at Boston to besiege Penobscot, of which but little notice was taken. Capt. Henry Mowat, of his Majesty s sloop *Albany*, having been many years on the American station and well acquainted with the disposition of the inhabitants, and of the importance of the country of Penobscot to the Americans for firewood, lumber, masts, cod and river fish, gave credit to the information, and ordered the three sloops of war into the best situation to defend the harbour, annoy the Enemy and co-operate with the land forces.

* The 74th Foot, "The Argyle Highlanders/ was raised by the Duke of Hamilton and served in America four years, under John Campbell. Milltown, New Brunswick, and a tract of good farming land on the Digdequash, were granted to the officers and men of this regiment who had been) in the garrison on the Penobscot.

** The 82nd Foot, " The Hamilton Regiment," served in America four years and was under Sir William Erskine in 1779- After the completion of Fort George, at Castine, this regiment returned to Halifax with General McLean.

Sir John Moore, made famous by Wolfe's poem on his burial in 1809, was then but eighteen, a

141

lieutenant, as was Sir James Craig, who became the Governor General of Canada. ,

July 19. The intelligence of yesterday gains credit; whereupon the General, in order to make the proper dispositions for an Immediate defence, desists for the present from his purpose of proceeding in a regular way with the fort; and prepares to fortify in a manner more expeditious and better suited to the present emergency; in doing which he shows the utmost vigilance and activity, giving every where the necessary directions, visiting incessantly by night and day the different parts of the works, and thus by his example animating his men to proceed, regardless of fatigue, with vigour and alacrity in their operations. The Inspector of the inhabitants begs leave of the General to call in the people to assist in carrying on the works ; which being granted, about a hundred inhabitants came in (with their Captain* at their head) as volunteers; and having worked three days gratis, cleared the land of wood in the front of the fort, to the satisfaction of the General, who returned them his thanks.

July 20. All hands busy at work, preparing to receive the enemy. At noon Capt. Mowat, having made every preparation in his power to secure the harbour, &, sent 180 men on shore from the ships of war, to work on the fort.

July 21. Intelligence is received that a fleet of near 40 sail of vessel had sailed from Boston eastward. All hands at work day and night. July 22. Nothing remarkable. All hands at work day and night. This evening a spy brought an account that 40 sail of vessel put into Townsend Harbour yesterday.

July 23. Every person busily employed. The Inspector calls a great number of inhabitants to work, who are employed in felling trees, raising an abatis round the fort, building plat forms for the guns, &c. Saw three sail in the offing. Several canoes from the islands below come to advise the

142

General of a large number of vessels being becalmed off St. George's Island,* standing with their heads to the eastward. All doubt of an attack from the Enemy is now vanished.

* The St. George Islands are off the mouth of St. Georges River and are a part of the town of St. George,, Maine. They were originally a part of the Plantation of St. George, then dishing, and in 1903 the town of St. George was incorporated. Penobscot Harbor, referred to by Rosier, in 1605, is at Allen s Island and here was the first attempt by Europeans to cultivate the soil of Maine. Captain George Weymouth erected a cross on Allen's Island in 1605, and the Maine Historical Society erected a granite one in 1Q05 to commemorate the event. The town of St. George is thirteen miles south of Rockland.

July 24. At 4 P. M. discovered a large fleet standing up the bay, which from various circumstances we believed to be the armament that, according to intelligence received, had been fitted out at Boston to besiege this place. On this account Capt. Mowat thought proper to detain the North and Nautilus sloops, which had been ordered for other service. At five, by signal from the Albany, the seamen who for some days past had been at work raising the S. E. bastion of the fort, repaired on board their respective ships (which were immediately cleared for action) and, as had been usual, were every evening exercised at their quarters. The Albany, North and Nautilus had dropped down the harbour and moored in a well-formed and close line of battle across the entrance, immediately within the rocks on Bagwaduce point and the point of Nautilus or Cross Island; giving a berth, out of the line of fire, to three transports stationed and prepared to slip and run foul of the Enemy s ships, should they attempt to enter the harbour. The troops were encamped about half a mile from the works; the well bastion of which was not yet begun, nor the

Seamen's** quite finished.

** So called as being the work of the seamen only under the direction of Lieut. William Brooke, of his Majesty's ship *North*. were put in a more defensible state, some cannon were mounted, and the little army was in garrison early the next morning. Guard-boats, during the night, watched the motions of the Enemy, who were discovered to have come to an anchor about three or four leagues off, in the narrows of Penobscot.

July 25. At 10 A. M. a brig appeared at some distance from the harbour s mouth, and after reconnoitering the situation of the men of war, stood back into the fleet. At noon the Enemy's fleet, consisting of 37 sail of ships, brigs and transports, arrived in the bay of the harbour ; the transports proceeded about half a mile up Penobscot river, and came to an anchor, while the armed ships and brigs stood off and on and a boat from each ship repaired on board their flagship, which had thrown thrown out a signal for that purpose.

At 3 P. M., nine ships, forming into three divisions, stood towards the King s ships and, as they advanced in the line, hove-to, and engaged. A very brisk cannonade continued four glasses*, when the Enemy bore up, and came to an anchor in the bay without. The King s ships suffered only in their rigging. The fire of the Enemy was random and irregular, and their manoevres, as to backing and filling, bespoke confusion, particularly in the first division, which scarcely got from the line of fire when the second began to engage. The second and third divisions appeared to have but one object in view, that of cutting the springs of the men of war, to swing them from the bearings of their broadsides, and thereby to afford their fleet an entrance into the harbour. During the cannonade with the shipping the Enemy made an attempt to land their troops on Bagwaduce, but

144

were repulsed with some loss. On the retreat of the Enemy s troops and ships the garrison manned their works, and gave three cheers to the men-of-war, which were returned; and soon after the general and field-officers went down to the beach and also gave three cheers, which were returned by the ships. Guard-boats and ships companies during the night lay at their quarters.

* "glass" is a marine measure of time, equal to half an hour.

July 26. At 10 A. M. the Enemy's ships got under weigh, and forming their divisions as yesterday, stood in and engaged the King s ships four glasses and a half. The damages sustained this day, also, were chiefly in the rigging at the extreme ends of the ships; and the fire of the Enemy appears again to be directed to the moorings; which attempt not proving successful, they bore up and anchored without. The Enemy again attempted to land their troops, but were driven back ith some little loss. At 6 P. M. the Enemy, having stationed two brigs of 14 guns and one sloop of 12, on the east side of Nautilus Island, landed 200 men, and dislodging a party of 20 marines, took possession of four 4-pounders (two not mounted) and a small quantity of ammunition. At 9 P. M. it being found that the Enemy were very busy at work, and that they had landed some heavy artillery which they were getting e men-of-war could not act in their present station, it was judged expedient to move them farther up the river. This was accordingly done, and the line formed as before: the transports moved up at the same time and anchored within the men-of-war. Guard-boats and the ships companies, as usual, lying at their quarters.

July 27. Pretty quiet all this day. A few shots from some ships of the Enemy were aimed at the small battery on Majabigwaduce point, which were returned with a degree of success, one ship having been driven from her station. Observed the Enemy very busy in erecting their battery on Nautilus Island. The garrison being much in want of cannon, some guns from the transports and from the off-side of the men-of-war, were landed, and being dragged by the seamen up to the fort, were disposed of for its use. At 3 P. M. a boat passing from the Enemy's ships to Nautilus island was sunk by a random shot from the fort. At 11 P. M. the guard-boats from the King's ships fell in and exchanged a few shots with the Enemys.

July 28. At 3 A. M. under their ships fire, the Enemy made good their landing on Majabigwaduce, and from their great superiority of numbers obliged the King's troops to retreat to the garrison. The Enemy s right pressed hard and in force upon the left of the King s troops, and attempted to cut off a party of men at the small battery; but the judgment and experience of a brave officer (Lieut. Caffrac, of the 82nd) counteracted their designs, and a retreat was effected with all the order and regularity necessary on such occasions. An attempt was made to demolish the guns, but the Enemy pushed their force to this ground so rapidly as not to suffer it. The possession of this battery afforded their ships a nearer station, on which they immediately seized.

At 6 A. M. the Enemy opened their battery of 18 and 12 pounders from Nautilus is land, and kept up the whole day a brisk and well-directed fire against the men-of-war. The King's ships cannonaded the battery for two glasses, and killed some men at it; but their light metal (six pounders) was found to be of little service, in comparison to the damages they sustained from such heavy metal brought against them. At 10 A. M., 13, of 32 guns, the Commodore's ship had not as yet been in action, got under weigh and with three more ships showed an appearance of entering the harbour, but hauled by the wind at a long distance.

146

A brisk fire was kept up for half an hour, when the Enemy bore up and came to an anchor again without. The *Warren* suffered considerably: her mainmast shot through in two places, the gammoning of her bowsprit cut to pieces, and her forestay shot away. Their confusion appeared to be great, and very nearly occasioned her getting on shore, so that they were obliged to let go an anchor and drop into the inlet between Majabigwaduce head and the point; where the ship lay this and the next day repairing her damages. The battery on the island still keeping up a heavy fire, and the ships crews being exposed without the least benefit to the service, Capt. Mowat thought proper to move further up the harbour; which was done in the night and the line formed again; he being firmly resolved to dispute the harbour to the last extremity, as on that entirely depended the safety of the garrison, whose communication with the men-of-war was of the utmost importance.

The dispositions on shore and on the water co-operating, and perfectly supporting each other, foiled the Enemy in their purposes; their troops were yet confined to a spot they could not move from, and while the harbour was secure their intentions of making approaches and investing the fort on all sides could by no means be put in execution. The present station of the men-of-war being such as rendered it impossible for the Enemy's ships to act but at particular periods, the marines (whose service in their peculiar line of duty was not immediately required on board) were ordered on shore to garrison duty, holding themselves in readiness to embark at a moment s notice, which with ease they could have effected in ten or fifteen minutes. Guard boats as usual during the night.

July 29. At 6 A. M. the Enemy s ships weighed, and altering their positions, came to an anchor again. The State of the fortress requiring more cannon, some remaining off-side guns were landed from

the men-of-war and dragged by the seamen up to
the fortress for its use and that of the batteries;
and though the task to be performed, up a steep
hill, over rocks and innumerable stumps of fallen
trees, was laborious, yet their chearfulness and
zeal for the service surmounted every difficulty.

P. M. the Enemy opened their batteries on the heights of
Majabigwaduce, and kept up a warm and
incessant fire against the fortress. The
commanding ground of the Enemy s works and
the short distance from the fortress, gave them
some advantages with their grape as well as round
shot which considerably damaged the storehouse
in the garrison. Six pieces of cannon at the half-
moon battery near Banks house, and which
belonged to the fortress, being now found
necessary for its particular defence, were moved
up to it and replaced with some ship s guns,
under the direction of the gunner of the Albany,
with a party of seamen Capt. Mowat having
obtained intelligence that the Enemy, in despair of
reducing the Kings ships by the means of their
own, or of getting possession of the harbour, had
come to the resolution of joining their whole force
in troops, marines and seamen, to storm the
fortress the next morning at daybreak, he judged it
expedient to reinforce the garrison with what
seamen could be conveniently spared; and for this
purpose, at the close of the evening, 140 men
under the command of Lieut. Brooke, were sent
into garrison: part of them were immediately
detached to re-inforce the troops on the out-line
piquets, others manned the facing of their own
bastion, while the remainder were busily employed
in raising the cavaliers in the fort. In all these
operations a brotherly affection appeared to unite
the forces both by sea and land, and to direct
their views all to one point, much to their credit
and to the honour and benefit of the service.
During the night the Enemy threw a number of
shells into the fortress. At 10 P. M. a few shot

between the Enemy s guard-boats and those from the Kings ships.

July 30. The Enemy s ships preserve their disposition of yesterday. A brisk cannonade the whole day between the fortress and the Enemy s batteries on the height, and a number of shells thrown on both sides. The storehouse being apprehended to be in danger, some seamen were ordered to move the provisions out of the fortress into the ditch in its rear; as likewise a quantity at another storehouse. Guard-boats as usual.

July 31. At 2 A. M. the seamen and marines of the Enemys fleet landed to the westward of the half-moon battery, and under cover of the night attacked the piquet, and by heavy platoon firings obliged them to retreat; but an alert re-inforcement of 50 men who were detached from the garrison, under the command of Lieut. John Graham of the 82nd regiment, to the support of the piquet, drove the Enemy back with some loss in killed, wounded and taken, amounting on the whole, according to the best information, to about 100; the loss on the part of the King s forces, amounting to 13 killed, wounded and missing, fell chiefly on the seamen and marines, who com posed the piquet this night. Lieut. Graham unfortunately received a dangerous wound in this action.

August 1. A slack fire on all sides. At 4 P. M. the Enemys fleet getting under weigh, and the wind and tide serving them to enter the harbour, the embodied seamen were immediately called on board their respective ships; but it afterward appeared that the Enemy weighed only to form a closer line. Guard-boats as usual.

August 2. At 10 A. M. three of the Enemy s ships weighed and came to an anchor nearer the harbour s mouth. Some can nonading between the fortress and the Enemy s batteries on the height. The outer magazine of the fortress being

149

too much exposed, as lying in front and between
the two fires, the marines were charged with the
duty of bringing it to the magazine in the fortress;
which was performed without any loss. P. M. a flag
of truce from the Enemy, to treat for the
exchange of a lieutenant of their fleet taken
(wounded) at the half⁻moon battery on the 31st
ult., but he had died of his wounds this morning.
TJiis day the Enemy posted some marksmen
behind trees within musquet⁻shot of the fortress,
and killed and wounded some centinels.

August 3. A slack fire the whole day. Perceived the
Enemy busy in erecting a battery to the northward
on the main above the King s ships. By a deserter
from the Enemy s fleet we learn the force landed
below the half ⁻moon battery was 1000 seamen
and marines, joined on their landing by 200
troops; that their intentions were to storm the
fortress in the rear while the army from the
heights made their attack in front; that it was not
intended to storm the half ⁻moon battery, but that
they had mistaken their road in endeavoring to get
in the rear of the fortress, when they received the
first fire of the piquet, which led them to suppose
their design had been dis covered, and that they
were ambushed. The army also, believing this to
be the case, retreated to their ground.

At 2 P. M. some seamen were sent to the fortress to
assist in working the cannon, and another party for
the defence of the Seamens bastion, where a
number of swivels from the men⁻of⁻war were
planted, loaded with grape⁻shot, as a precaution
against any attempt of the Enemy to storm the
works. By request of the General a number of
pikes were also brought from the Kings ships to
the fortress, and put in the hands of the seamen,
to prevent the Enemy from BOARDING their
bastion. Guard⁻boats as usual.

August 4. The Enemy s ships retain their former
situation. A smart cannonading between the

150

fortress and the batteries on the heights, and a great number of shells thrown on both sides. Some ships buckets for the use of the garrison brought on shore, in case the fascines at the well bastion, or storehouses might be fired by the Enemy s shells. At 9 A. M. the Enemy opened their new battery near Wescoat s house, on the main, to the northward of the shipping. A brisk fire was kept up the whole day, and the men-of-war suffered much in their hulls and rigging; being too far from the battery for the light metal of the ships to produce any effect, their companies were ordered below. P. M. some skirmishing between the piquets, and trifling losses on both sides, on the Enemy s some Indians were killed.

During the day several accidents happened by cannon shot in the fort; among others the boatswain of the Nautilus was wounded by grape, and a seaman belonging to the North killed by an 18-pounder, at the guns they were stationed at in the fortress.

August 5. Cannonading the greatest part of the day between the fortress and the Enemy s batteries on the height, and from the north battery against the men-of-war, damaging their hulls and rigging. A. M. the remaining off-side guns from his Majestys sloop North brought on shore, and mounted in the cavalier in the fortress. P. M. the garrison, being much in want of wads and match, was supplied from the men-of-war, as also with some six-pound shot, in which it is deficient. The north battery on the main having the command of the opposite shore on the peninsula of Majabigwaduce, where the Enemy, under its protection, might make lodgements in their approaches toward the heights opposite the men-of-war and within shot of the fortress, and might thereby destroy the communication between them and the garrison, Capt. Mowat judged it necessary to erect a work in order to preserve this communication: a square redoubt was therefore marked out, to be manned

151

with 50 seamen and to mount eight ships guns en barbette. Guard-boats as usual during the night.

August 6. Slack fire between the fortress and batteries on the heights, and a few shot from the north battery against the men-of-war, cutting their rigging and dismounting a six-pounder on board the North. At 4 A. M. 70 seamen from the different ships, under the direction of Lieut. Brooke, of the *North*; sent on shore to raise the Seamen's redoubt on the height. P. M. a quantity of musquet-cartridges (of which the garrison was in want) brought on shore from the men-of-war. Guard-boats as usual. At 11 a few shot exchanged between the guard-boats.

August 7. The Enemy s ships preserve their positions. At 9 A. M. three of their brigs got under weigh and stood down the bay, supposed on the lookout. Some skirmishing between the piquets, with loss to the Enemy; Lieut. McNeil,* of the 82d, and one private, wounded. Slack fire between the batteries and the fortress, and the north battery perfectly silent. At 4 P. M. discovered a boat crossing the S. E. bay to Haineys plantation, where the Enemy kept a piquet. Lieut. Congalton,t of the *Nautilus* chaced with the boats from the men-of-war, and took her ; but her crew, with those of a whale-boat and a gondola for transportating cannon, got safe oil shore and joined the piquet. Capt. Farnham J of the *Nautilus,* with Lieut. Brooke and 50 seamen, joined by a party of soldiers from the garrison, landed and scoured the woods; the Enemy fled immediately, and so effectually concealed themselves as not to be discovered; some had left their arms ammunition and blankets, which were taken and brought on board.

Guard-boats as usual during the night.

By a deserter from the Enemy we learn that General Lovell had sent out small parties from his army, round the country, and brought in a great

152

number of loyal inhabitants, who were sent on board their fleet and thrust down the holds heavily laden with irons, both on the hands and feet; their milch cows and other stock killed for the Enemy s use; all their moveables destroyed or plundered, and their wives and children left destitute of every support of life.

August 8. A constant cannonade the whole day between the fortress and the Enemy s batteries on the height, and from the north battery against the men-of-war, but returned only with a musquet. At 10 A. M. the Enemy brought a field-piece to play from the main on the seamen working at the redoubt; but the facing towards the Enemy being the first raised, for the purpose of covering the party, it was impossible to dislodge them; and a covering party daily attending from the garrison prevented a nearer approach on any other ground. This evening the redoubt was finished, and to the credit of the seamen, met with the approbation of the General and Engineers.

Guard-boats as usual.

August 9. Cannonading as usual. At 9 A. M. a new battery, on the left of the Enemy s lines, was opened against the fortrees, and its chief fire, as well as the shells, directed against the N. W. bastion, raised with fascines only. P. M. discovered the Enemy had moved their piquet from Haineys plantation, and given up their design of carrying on a work for two 18-pounders against the men-of-war.

Guard-boats as usual during the night.

August 10. The Enemy s ships in the former position. A slack fire on all sides, and nothing material.

August. 11. A smart cannonading from all the batteries, and some shot from the north battery well directed at the men-of war.

August 12. Slack fire on all sides, and no material operations the whole day; but at 9 P. M. a large body of seamen and marines from the Enemy's fleet landed below Banks* house to the westward, and setting fire to some barns, houses, and a quantity of lumber-boards, &c., on the beach, retreated to their ships again.

August 13. At day break some skirmishing between the piquets, but no material loss on either side. At 1 P. M. came in some deserters from the Enemy s ships, who say the boat chaced on shore at Hainey s plantation had in her their Commodore and some officers of their fleet, who, having escaped, returned to their ships after lying two days and a night in the woods; that one of the officers (Capt. Ross, of the Monmouth) had broke his leg in the woods; and that they were much disconcerted at the loss of the gondola, which was intended to carry over some 18-pounders to the battery on the plantation.

Capt. Mowat also (by his usual diligence) obtained information that a degree of mutiny prevailed in the Enemy's fleet against their Commodore who,notwithstanding the resolves of several councils of war and urgent solicitations of the General to make another attempt on the Kings ships, had hitherto declined it through fear of losing some ships; but that, in consequence of another council held this morning on the Warren, it was determined to force the harbour next tide and take or destroy the men-of-war; that five ships were destined for this service, one of which was the *Warren;* but that the *Putnam*, of 20 guns, was to lead, and that each ship was doubly manned with picked men. This information was confirmed at noon by five c/f their fleet getting under weigh and coming to an anchor in a line, the *Putnam* being the headmost ship. The marines were now called on board their respective ships, the barricades strengthened, guns double-shotted and every disposition made for the most vigorous defence. The *St. Helena* transport had been brought into the line and fitted out with what guns

154

could be procured, and the crews of the transports (now scuttled and laid on shore to prevent them from falling into the Enemy s hands), turned on board to fight her; and the General had also advanced five pieces of cannon, under cover of an epaulement, to salute them as they came in.

*The home of Aaron Banks, a soldier of the French and Indian wars, who came from York, Maine, in 1765. He married Mary Perkins of York, who was a sister of John and Daniel Perkins of Bagaduce. He died August 9, 1823, at Penobscot, where he moved after peace was declared. He has no descendants of his name. Banks and his family were detained for upwards of three weeks as prisoners on board the British sloop North.

But at 5 P. M. the appearance of some strange sails in the offing disconcerted the Enemys plan, and the five ships, getting under weigh again, stood off and on the whole night. Guardboats watching the motions of the Enemys fleet, and the ships companies standing at their quarters until daylight. This night had been fixed upon to storm the north battery with 60 seamen under the command of Lieut. Brooke, supported by Lieut. Caffrac of the 82d, with 50 soldiers; but the Enemys operations, and the appearance of the strange fleet, prevented the execution of it.

August 14. At daybreak this morning it was discovered that the Enemy had during the night moved off their cannon, and quitting the heights of Majabigwaduce, silently embarked in small vessels. At 4 A. M. after firing a shot or two, they also evacuated Nautilus island; and leaving their cannon spiked and dismounted, got on board a brig lying to receive them, and made sail with the transports up Penobscot river. The whole fleet now got under weigh, and upon one of the brigs heaving in sight off the harbours mouth, with various signals aboard, they bore up with all sail after the transports. There now remaining no

155

doubt but the strange fleet was the relief expected, the off-side guns of the Albany, North and Nautilus were got down from the fortress, and being taken on board, the three ships slipped their stern moorings, hove up their bower anchors, and working out of the harbour joined in about the centre of the King s fleet, in pursuit of the flying enemy, who were now crowding with every sail they could set. The Hunter and Hampden, two of the Enemy s ships, of 20 guns each, attempted to escape through the passage of Long Island,* but were cut off and taken; the former ran in shore all standing, and was instantly deserted by her crew, who got safe on shore; and the *Raisonable*, Sir George Collier, being the sternmost ship in the fleet, took possession and got her off, and came to an anchor near her. The rest of his Majestys ships continued in chace of the Enemy until it grew so dark as to render the narrow navigation exceedingly dangerous; and they were obliged to anchor for the night, while the Enemy, having good pilots, ran some miles further up the river. The *Defiance* brig, of 14 guns, ran into an inlet where she could not be pursued, and was set on fire by her crew. During the night the Enemy set fire to several ships and brigs, which blew up with vast explosions.

* Long Island, now the town of Islesborough, is about twelve miles long, Contains about six thousand acres and is in Penobscot Bay, four miles from Castine.

In short, the harmony and good understanding that subsisted amongst the forces by by sea and by land enabled them to effect almost prodigies; for so ardently did they vie with each other in the general service that it may be truly said not a single OfficerSailor or Soldier was once seen to shrink from his duty, difficult and hazardous as it was. The flying scout, of 50 men commanded by Lieut. Caff rac of the 82d, in particular distinguished themselves to admiration, marching frequently almost round the peninsula, both by day and by night, and with drum and fife playing the tune called Yankee, which greatly dispirited the Enemy, and prevented their small parties from galling our men at the works. In one instance they even drove back to their incampment 300 of the Enemy who had been sent to storm an outwork.

The manoeuvres of the three Sloops of War, under the direction of Capt. Mowat, were moreover such as enabled the King s forces to hold out a close siege of 21 days, against a fleet and army of more than six times their number and strength; insomuch that on the first appearance of the re⁻in⁻ forcement from New York in the offing, the Enemy debarked their troops and sailed with their whole fleet up Penobscot river, where they burnt their shipping and from thence marched to their respective homes; and the loyal inhabitants, who were taken in the time of the siege and cruelly treated on board their ships, had their irons taken off and were set at liberty.*

Thus did this little Garrison,** with three Sloops of War,; in an enterprise of great importance, against difficulties apparently insurmountable, under circumstances exceedingly critical, and in a manner strongly expressive of their faithful and spirited attachment to the interests of their King and Country.

*To give them a cool airing, as the enemy called it, once a day the irons were knocked off their feet and they were put into a boat alongside the ship, where they remained about an hour, and had the filth of the ship poured upon their heads.

**When the account of an army coming to besiege this place was received, the curtains in some parts; of the intended fort were not more than four feet in height; two bastions were but just begun to be built, and the other two were only marked out.

A LIST of the Enemy s Ships, etc., taken and destroyed in Penobscot River by the unwearied exertions of Soldiers and Seamen whose bravery cannot be too much extolled, under the judicious conduct of Officers whose zeal is hardly to be paralleled, succeed

Ship's Name	Commanders	Guns	# Men	Metal Pounders	
Warren	Saltonstall	32	250	8 and 12	Burnt
Sally	Holmes	22	200	9 and 6	Burnt
Putnam	Waters	20	130	9	Burnt
Hector	Cairns	20	130	9	Burnt
Revenge	Hallet	20	120	6	Burnt
Monmouth	Ross	20	100	6	Burnt
Hampden	Salter	20	130	9 and 6	Taken
Hunter	Brown	20	130	6	Taken
Vengeance	Thomas	18	140	9 and 6	Burnt
Black Prince	West	18	100	6	Burnt
Sky Rocket	Burke	16	120	6	Burnt
Brigs					
Hazard	Williams	18	100	6	Burnt
Active		16	100	6	Burnt
Tyrannicide	Cathcart	14	90	6	Burnt
Defiance		14	90	6	Burnt
Diligence	Brown	14	+90	4	Burnt
Pallas	Johnstone	14	80	4	Burnt
Sloop Providence	Hacker	12	50	6	Burnt
Nine Transport Vessels			Taken	2	
Ten Transport and Ordnance ditto			Burnt	35	
Total				37	

Killed, wounded and missing, of his Majesty's Sea and Land Forces 70

Killed, wounded and taken, on the Enemy's Side 474

Of the captains of these vessels the Massachusetts records show particulars. The Sally is described as the Charming Sally, a privateer owned by William Erskine of Boston. Captain Alexander Holmes was afterwards captain of the privateer Batchelor.

158

William Burke commanded the Skyrocket, which
   was a privateer owned by Ebenezer Parsons,
   of Boston.

James Johnston was the captain of the Pallas^
   privateer, owned by William Erskine and
   others, of Boston.

Nathan Brown, of Salem, commanded the Hunter,
   a privateer owned by Bartholomew Putnam.
   Later he had the privateer-ship Jack.

John Cathcart, captain of the Tyrannicide,
   afterward had command of the State ship
   Tartar and another of the same name, a
   Boston privateer.

John Games (not Cairns) had the ship Hector, a
   Boston privateer owned by Jonathan Peale,
   and afterward of the Montgomery and Porus,
   both privateers.

Allen (or John Allen) Hallet, of the Active, a State
   vessel, was afterward in command of the
   Tartar and the Franklin and Minerva,
   privateers.

Captain Hoysteed Hacker commanded the
   Providence and afterward the privateer ship
   Bucanier.

Nathanel West was captain of the Black Prince,
   privateer owned by George Williams, of
   Salem. He afterwards had the Three Sisters,
   owned by Nathaniel Silsbee and Elias Hasket
   Derby of Salem, and of the Marquis.

Daniel Waters was captain of the General Putnam,
   which was owned by Nathaniel Shaw. He
   had previously commanded the Lee and
   afterwards had the Friendship.

CAPTAIN HENRY MOWAT'S ACCOUNT
of the occupancy of the Penobscot by the British during the Revolution

In the catalogue of a London bookseller, in 1843, appeared for sale a manuscript relating to the services of Capt. Henry Mowat in America. It was disposed of, to whom was unknown. The title was " *A relation of the services in which Captain Henry Mowat was engaged in America, from 1759 to the end of the American War in 1783.*"

Search was instituted by Maine historians for the manuscript. Judge Joseph Williamson, of Belfast, Maine, advertised abroad, in 1887, " *I will pay five pounds for evidence of the existence of the manuscript.*" On November 20, 1890, it was received by Hon. James P. Baxter, of Portland, from Edinburgh, and was published in part, with its history, in the Collections of the Maine Historical Society, Series II, Vol. 2, page 345. The original manuscript of fifty-nine pages is now in the possession of that society.

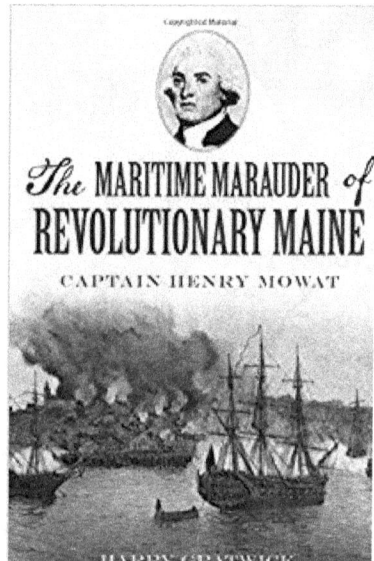

The following is the part relating to the occupancy of the Penobscot by the British during the Revolution; beginning with the middle of page 7 and ending near the bottom of page 21. Punctuation, spelling and capitalization are as in the original.)

Excerpts from Mowat's Manuscript

THE Albany at last was called to New York in the beginning of 1779 orders had not long before arrived from Britain for taking post in Penobscot Bay, and Capt. Mowat's experience of the New England Coast being well known to Sir Henry Clinton on former occasions, he was proposed by his Excellency approved by Admiral Gambier as the fittest to command the naval part of the Force. The Admiral desiring to know the force necessary for the Service, was answered it should be Superior to any the Enemy at Boston could readily collect on such Emergency. It was accordingly settled it should be so, and that Captain Mowat should have a ship equal to the Importance of the object.

In the meantime the Store of Powder in the Garrison at Halifax being totally exhausted, Captain Mowat received on board the Albany and proceeded with an ample Supply, the orders and Every equipment for the Expedition, being intended to follow: but he had no sooner landed the Powder, than he was ordered by Sir George Collier to the Bay of Fundy, and Sir George repaired soon after to New York where he was left the Senior Officer on the American Station.

On this change taking place, Captain Mowat, from reasons otherwise foreign to this Narrative, Considered it Necessary to urge what he had formerly represented to Admiral Gambier, and he wrote to New York from the Bay of Fundy, that if the Albany were to be the leading Ship, it would by no means be safe to trust the Expedition with one of her class, unless a Sufficient force should cruize between it & the enemy, until the post should be established.

161

This representation appears to have had no effect, for the orders for the Albany alone soon after arrived at Halifax, and were delivered by Capt. Gaylor of the Romulus to General McLean until the Albany should arrive.

Thus, if the Albany had happened to lead the Expedition according to the order, the whole must have been intercepted as we shall shortly see, & carried to Boston for a mere Novice might have conceived at once She was not fit to conduct it safely. The Consequences, which must be estimated according to the view & State of affairs at that time in America, Would have been tremendous. It would have been equivalent to a Second Burgoynade before there were time for repairing, or forgetting, the first: an immense Encouragement for the Americans, who were tiring of the length of the war, to exert their remaining resources, for the Opposition to exercise their clamor, and a proportional depression of the Spirits of the Loyalists. To the Southward we had but a slender footing in Georgia against such a disaster, the reinforce ments not arrived as yet And the Army there inactive for Security. To the Northward Canada was not so strong as it had been rendered in the Succeeding Year, And Nova Scotia at least, lying contiguous to the territory of Penobscot, would have been over whelmed, for by this detachment the Garrison of Halifax had been by the one-half reduced. This disposition of the Service must appear the more strange as we know Sir George Collier was by no means ignorant of the rebel force in the New England Ports.

But the dire Event was prevented by a mere accident & that the most fortunate in the World; for the Dispatch, forwarded by General McLean, did not reach the Bay of Fundy where Capt. Mowat was stationed, nor did he in Consequence get round to

Halifax, until the latest moment having elapsed the General put the order into the hands of Captain Barclay* of the Blonde Frigate, then Senior officer of the Navy there, who immediately put the North & Nautilus sloops of war under orders to proceed with himself And they were on the point of sailing when the Albany arrived. However this did not alter Captain Barclay's judicious determination. They proceeded, had a long passage As might be expected at the Season, and at last arrived at Penobscot: The Rebel frigates. Boston & Providence, who were cruizing on the Coast of Nova Scotia westward of Halifax, finding the Convoy Superior to what they expected, did not think proper to attack it.

In a few days after the troops were landed, the Blonde departed, leaving Captain Mowat under a copy of Sir George Collier s original orders, with directions for the North and Nautilus & all the transports to return to Halifax. Now soon the stores were landed for Captain Barclay had brought the Sloops of War there without Sir George Collier s orders, Captain Mowat finding the wretched Albany was to be left thus alone, to lie in an open harbour distant from every Aid, and in the Jaws of the most powerful of the rebellious Colonies, to co-operate with about 700 troops in a fort not yet begun to be erected, was convinced it would be for the good of His Majesty s Service to use the utmost Latitude, the order would admit of, to postpone the departure of the Ships, from the following view of the Situation of the Armament.

* Andrew Barklay, the captain of the
frigate Blonde, called by one who saw her,
" a beautiful ship/ was a Loyalist from
Boston. He was a protestor against the
Whigs in 1774. After peace was declared,
accompanied by his family of ten persons
and by four servants, he left New York for

Shelburne, Nova Scotia, where the Crown granted him fifty acres of land, one town and one water lot. He was living there in 1805.

The Bay of the Penobscot is spacious and capable of containing all the Navy in the World. In a corner of it about fourteen leagues distant from the open Sea, near the Embrochure (sic) of Penobscot River is the Harbour of Magebigwaduce. This Harbour is formed on the one Side by the Mainland, and along the entire other side of it Stretches the Peninsula of Magebigwaduce. Cross* (now Nautilus Island) is at the entrance of the Harbor. The Peninsula of Magebigwaduce is a high Ridge of land at that time much encumbered with wood. To its summit, where the fort was ordered to be erected there is an ascent of more than a quarter of a mile from the nearest shore of the harbour.

The Provisions, Artillery and Engineer Stores and the equip age of the troops, being landed on the Beach, must be carried to the Ground of the fort chiefly by the labor of the men against the as ent, there being only a Couple of small teams to Assist in it. The ground & all the Avenues to it, was to be examined, cleared from wood, and at the same time guarded. Materials were to be collected & prepared, And the defences, as well as every convenience of the fort, were to be reared. Let any one conversant in Matters of this Nature, reflect what a work it was for 700 men, And he will also readily allow, that in the Course of it they could not possibly, whether from fatigue, or in point of Necessary Preparation be in Condition of repelling any powerful attack. That, as appears also from the rebel General Level s letter, everything depended on our Men of War being able to prevent the Enemy from entering the Harbour, which was not liable to be commanded or protected by the Guns of the Fort. That the Harbour once forced, a Superior Number of the Enemy might land on the most convenient parts of the Peninsula, cut off the communication of our Troops with that considerable part of the Necessary Stores, which to the last while the fort was erecting, must unavoidably be left on the Beach, force them to retire within the unfinished Breastwork, where Surrounded without cover, Comfort or defence, they could have no alternative but to yield Prisoners of War in a few days, or to risk an action against thrice their number on ground from its Nature more favor able to the Enemy s mode of fighting than for theirs. It is altogether Superfluous to comment any farther on the orders by which a harbour, of this Importance must be left to the sole protection of the Albany Sloop, carrying ten Six and Six four pounders.

> * Nautilus or Cross Island, sometimes called Banks Island, for its owner, is southeast of Castine in Penobscot Bay and was named for the sloop of war Nautilus.

The *Blonde* Frigate had not been many days departed, when Captain Mowat having taken Measures for procuring the best information from Boston, concluded that the Post would soon be attacked, and he proposed to General McLean to give his concurrence for detaining the North & Nautilus, as well as the Transports, judging the General s Consent to be eligible, because otherwise he would be liable to Account for acting contrary to the orders left with him.

The General equally confident in the Intelligence, gave his Concurrence, and accordingly in the fifth week from the Arrival of the Royal Armament at Penobscot, the Rebel fleet appeared in the Bay, consisting of eighteen vessels of war as per the margin,** besides Transports having on board all necessary Stores and between two and three thousand Land forces.

** No list attached.

At that time a great portion of the stores had not as yet been carried up to the fort. Its Scite was lower by several feet, than a piece of ground at the distance of six hundred yards. The Parapet, fronting this higher ground was scarcely four feet high. All the other parts of the Parapet, paralell to the Harbour of Magebagwaduce and in the rear, were not three feet high. The two Bastions to the harbour were quite open. The troops were encamped on the area, which might be about the Space of an Acre, there had been a Shade erected for the Provisions. The Powder was lodged in covered holes dug in the proposed Glacis: There was but a Single Gun Mounted, & that a Six Pounder.

The Naval force in Magebagwaduce Harbour were the Albany, North & Nautilus, Sloops of War, Commanded by Captains Mowat, Selby and Farnham, and four Transports.

In this force and State of Preparation, one may easier conceive than describe the anxiety & hopes of all concerned on the appearance of so formidable an Armament.

The enemy came up, and paraded before the entrance of the harbour, in perfect confidence of entering it without difficulty, which would have been the case had the Albany been alone, and then everything would have been over at once; but there was such an excellent Disposition made of the Sloops of War & Transports in the entrance of the Harbour, as baffled every attempt of the Enemy to force it for three days then they prepared to land their troops on a Bluff of the Peninsula without the harbour, where the General could place pickets communicating with the Main body in the fort, to watch & to oppose, the debarkation.

These three or four days of Embarrassment on the part of the rebels gave our troops time to do something more to the Fort, to carry up the most necessary Stores, to mount several guns, and in short to devote every Endeavor to the present Exigency. The Enemy, having failed in their attempts on the harbour, effected at last a landing on the bluff, and by superior numbers forced the Pickets into the Fort, took possession of the high ground, above men tioned, within six hundred yards thereof & immediately erected their Batteries and Lines.

In this Position both Parties continued firing at one another during the whole Siege. Our Troops, tho extremely harassed, were daily getting into a better Situation, with the Assistance of the Seamen, and the Requisites which the Men of War furnished, as well as their own Stores. Secure on the Flanks & in the rear while our Ships maintained the Harbour, they had only to exert their chief attention & Efforts on the side fronting the Enemies Lines, which effectually deterred the latter from advancing in that direction.

They had erected Batteries on Nautilus Island, & in the rear of the harbour, all within point blanc shot shot of any position, in which the ships could be placed, but the proper choice of different stations on every emergency eluded their utmost efforts to enter it.

Thus both sides were employed, ashore & afloat, for 21 Days, in a variety of Manouveres, which are in part described in a Journal kept by an officer on shore & published by J. C. Esq.

In the Mean time Intelligence having reached New York, that Penobscot was attacked, Sir George

Collier Sailed to its relief, with the Raisonable Ship of the Line, Blonde, Virginia, Carmilla, Galatea, &c. They were perceived off Penobscot Bay by the rebel lookout vessel in the Evening. In the course of the night they embarked their Troops, &c., and in the Morning early their fleet was seen under Sail; but the wind failing them to get round the upper end of Long Island, they had no alternative but to run up Penobscot River. These Manouvres were a proof that the Strange Ships sailing up the Bay were a relief and the three Sloops of War being employed from daylight in embarking the part of their Guns that were ashore on the Batteries, &c., &c., were able to join in the center of the King s Ships: during the pursuit one of the rebel vessels struck, after a few shots, to the Blonde & Virginia: Another ran ashore at the same time some distance below the mouth of the River, and was some time after taken possession of by the Baisonable, which brought up the rear: All the rest, with the advantage of good pilots & of a whole flood tide which happened in the night, got such a distance up the River, as afforded time for destroying them, And the crews made the best of their way to New England, thro the woods, in the utmost distress.

Thus ended the attack on Penobscot. It was positively the severest blow received by the American Naval force during the War. The trade to Canada, which was intended, after the expected reduction of the Post of Penobscot, to be intercepted by this very armament, went safe that Season: The New England Provinces did not for the remaining period of the contest recover the loss of Ships, and the Expence of fitting out the Expedition: Every thought of attempting Canada, & Nova Scotia, was thenceforth laid aside, and the trade & Transports from the Banks of Newfoundland along the Coast of Nova Scotia, &c: enjoyed unusual Security.

168

After all was over, it was natural to be expected, that Sir George Collier would have been Supremely happy to have represented this important Service in its proper colors, and that Capt. Mowat would, according to the Custom of the Service, have been sent home with the Account: But in answer to the Claim, Sir George expressed the utmost regret, that he could not spare a Ship from the Station: assured that if he intended to send an officer to England Capt. Mowat would certainly be the person; that he only meant to transmit the Despatches by New York, in which he pledged his word, as he held it to be no more than his duty, that the Services of the Sloops of War would be represented in the most honorable Manner to the Admiralty

On the next day & before there was time to attend to writing the Official Account of the Siege, he put the Albany under orders to proceed up Penobscot River to the Rebel Wrecks, observing it would be some time before he would leave the Bay This done he departed abruptly for New York, and had no sooner gone out to Sea, than the Greyhound s Signal was made to part Company, And she proceeded directly to England with his Account.

Her destination had been Kept a Secret from everyone, General McLean excepted, who in his publick Letter Acknowledges having been privately informed. This is the Manner, in which Captain Mowat was prevented Sending an Official Account of the Siege, And, Notwithstanding Sir George Collier having solemnly pledged himself as above, we See his account to the Admiralty confined to the Merit which we will readily allow him of sailing from New York to the relief with a Squadron Which the United Naval force of All America was incompetent to resist even in a Crescent & to a description of the Disposition & destruction of the Rebel

Ships, which however could not be discerned by any one from on board the Raisondble: The Service of the three Sloops of War during the Siege were totally omitted & their Captains not even named.

When Admiral Arbuthnot s arrival had put an end to Sir George Collier s Command, Captain Mowat hoped some Justice would have been done him for the Service performed at Penobscot, at least so far as the laying a fair representation of it before the Admiralty, but there was not the least notice taken of him, and he was left at Magebigwaduce under a continuation of the distress of seeing also, that every Promotion, made by this Admiral, was without a single exception, of officers Junior to him: Among these an Officer, who had received his first Commission into the Albany when Captain Mowat was appointed to her, was made Post Captain: It is not from any individious (sic) Motive this Instance is given on Captain s Mowat's part : None can be more happy in the good fortune of an Officer, with whose great Merit he has had opportunities of being well Acquainted: but it is a Contrast to the glaring Injustice himself has Met with.

Henry Mowat was born in Scotland in 1734. He was the son of Capt. Patrick Mowat of H. M. S. Dolphin. After an experience of six years he was commissioned as lieutenant of the Baltimore in 1756. The certificate of his "passing" in the Admiralty records sets forth "He produceth records kept by himself in the Chesterfield and Ramlis (Ramillies) (as midshipman) and certificates from Captains Ogle and Hobbs of the Dilligence, etc.; he can splice, knots; reef a sail, etc., and is qualified to do the duty of an able seaman and midshipman." In 1762, he was promoted to be a commander and served as such on the Canceaux twelve years. It was during this time that he destroyed Falmouth Neck, now Portland. This event occurred October 18, 1775, and for it he was denounced by our forefathers and Washington wrote of his conduct, " I know not how sufficiently to detest it." Mowat was then forty-one years old. He had been captured at Falmouth Neck, the May before, and was released on his promise to return the next morning, which promise he did) not keep. His next vessel, the sloop Albany, was the flagship of the squadron at Penobscot. He served his King forty-four years, about thirty of which were on our coast. On board his ship, the Assistance, about five miles from Cape Henry, Va., April 14, 1798, he was stricken with apoplexy, died aged sixty-four years, and was buried in St. John's church yard, at Hampton, Va. He left a son, John Alexander, who entered the navy in 1804.

American

Journals, Proclamations and Letters

Written by those who where there

WILLIAM MOODY'S JOURNAL

William Moody of Falmouth, Maine kept a journal during his service in Col. Mitchell's regiment, recording each day the events that came under his observation. It is worthy of preservation. Mr. Moody was the drummer of Capt. Peter Warren's company. He had served in Col. Edmund Phinney's 31st regiment of foot at Cambridge in 1775, in Capt. Abner Lowell's Matross company at Falmouth Neck in 1776, 177, and 1778, and in Capt. Joseph Pride's company in Col. Joseph Prime's regiment at the same place in 1780. He was in the service in the early part of 1781, and went on a cruise in the privateer Fox, in April of that year.

Drummer of the day

Mr. Moody was the son of Enoch and Ann (Weeks) Moody of Falmouth, and was born February 16, 1756. He married Mary Young, in 1783, and had children Enoch, William and Nancy. He married for his second wife, Rachel Riggs in 1804, and had a son Edward. He died February 16, 1821, aged sixty-five years. His father, Enoch Moody, was the chairman of the committee at Falmouth in the Revolution, and his four brothers, Enoch Jr., Benjamin, Nathaniel and Lemuel, were Revolutionary soldiers.

The Journal

July 2. A detachment of 40 men to go to Major Bag a Duce.

July 3. Turned out as a Volunteer to go to Penobscot with Capt. Peter Warren.

July 9. Turned out in the morning for Exercise.

July 10. Our Regt. Paraded and arranged Capt. Warren's the first company.

July 14. The transports with 2 brigs & a sloop, a prize with 10 guns, arrived here to carry the Troops. Drew one day's allowance.

July 15. Drew 4 days' allowance.

July 16. Our Company embarked on board the sloop [Centurion] and hauled off, Capt. [William] McLellan master. [He was a son of Brice McLellan of Falmouth Neck.]

July 17. On shore to draw allowance and took it. Stayed all night.

July 19. I went on board of the Sloop Centurion [80 1/3 tons] at sunrise. Enbarked for Majibigwaduce. Weighed anchor at 8 oclock. Capt. [Abner] Lowell fired an 18 pounder for all hands on board. Arrived at Townsend [Boothbay] at 6 oclock.

July 20. Last night a soldier fired a gun and blowed his hand off, died. The Hampden a 20 gun ship arrived.

July 21. Went ashore to prayers. Parson [Thomas] Lancaster prayed and we sang. Between 30 & 40 sail of armed ships & Transports at Townsend.

July 22. Regt. Paraded ashore and Gen. [Solomon] Lovell reviewed them.

175

July 24. Admiral [Dudley Saltonstall] fired a gun about 4 oclock, the whole fleet under sail. Came to anchor at 9 o'clock at night under the Upper Fox Island.

July 25. Made sail for Bagaduce at 8 oclock. Came to anchor in Penobscot. The enemy fired from the shore with muskets. The ships ran in by the Forts and fired many broadsides. Seven of our boats that went to land got almost ashore. The enemy lay in ambush and fired upon us and killed an indian.

July 26. Our vessels warped in. We embarked our boats at 12 oclock. Kept off and on till sunset. [It is stated that the time was about 6 o'clock.] Come under the Admiral's [Frigate Warren] stern, then put off for an island [Nautilus] within point blank shot of the enemy's fort. As our boats were going across, the enemy sunk one boat by a (chain) shot and Major Daniel Littlefield [of Wells] and two others were drowned.

July 28. At daybreak had orders to land under cover of our guns on board the shipping. Commenced landing half an hour before sunrise. The enemy lay in ambush and firing upon us killed 1 capt. [probably Major Welch] of marines belonging to the Admiral and several others. We took 3 prisoners and killed 7. Have possession of the ground and soon hope to have all their works. 2 men wounded, one lost his leg and the other his arm. Went over to the Island after [Samuel] Knight. He was sick there.

July 29. The enemy throw shells. Loss and wounded in the attack [of 28th] about 30. Lost 1 man this afternoon.

July 30. Hauled up on the hill [over the high bluff where they landed] 2 eighteen pounders. A deserter came in from the enemy last night; he says the British force does not exceed 350. [This was not one half of the number of their men.]

July 31. Two seamen wounded with a shell who belongs to the Active. One of the marines belonging to the [frigate] Warren deserted to the enemy. Last night went out with a detachment of 88 men. Marched on to the parade at sunset and kept under arms till 2 o'clock [A.M.]. We then attacked one of the enemy's redoubts which we carried with the loss of a few men. We killed several of the enemy and took 18 prisoners. Capt. [Nathan] Merrill of our Regt. took one prisoner, a corporal of thee enemy.

Sunday, Aug. 1. Major [Samuel] Sawyer of the York [county] forces mortally wounded. He died this day.

Aug. 2. Mr. Wheeler Riggs [of Falmouth Neck] was killed this afternoon. One of the train badly wounded. Buried Mr. Riggs very decently. [He was stooping over fixing a gun carriage when a cannon ball hit a tree near, glanced and struck him on the back of his neck. He was the only Falmouth soldier killed in the expedition.]

Aug. 3. Gen. Lovell sent a flag to the lines to enquire after a Lieut. of Marines belonging to the *Vengence* who was missing after the battle of Sunday last [Aug. 1]. The answer returned was that the Lieut. was wounded in battle and died yesterday.

Aug. Wed. 4. Three of Capt. [Nehemiah) Curtis' men deserted. William Harper had a musket ball shot through his coat by the enemy while on picket guard.

Aug. 5. An Indian killed by the enemy, one taken prisoner. Capt. [David] Bradish from Falmouth to see us.

Aug. 6. Capt. Bradish and his crew left us. [He was sent to Boston.]

Aug. 7. Smart cannonading. Marched down towards the fort of the enemy about three o'clock. A party of about 100 sallied out. Gen. Lovell ordered a retreat to draw them out, but they immediately ran back to their entrenchment. One man belonging to Col. [Samuel] McCobb's Regt. Wounded.

Monday, Aug. 9. Attempted to land on Hyannis Point, opposite the enemy, but were prevented by the annoyance of the enemy in ambush.

Aug. 11. Last night [10th] 20 of Major (Nathaniel) Cousin's Regt. Deserted. One of the enemy deserted.

Aug. 12. Major Cousin's men brought back last night.

Aug. 13. Made another demonstration upon the lines of the enemy, but could not bring on an engagement. Capt. Woodman slightly wounded.

Aug. 14. News that the fleet of the enemy are at the mouth of the [Penobscot] Bay. We began our retreat about one o'clock. Ran with our Ships and Transports to Fort Penobscot and called on the Commissary for provisions. The enemy in sight and under cloud of sail. Some of our Ships are taken and some are run ashore. I took the boats and went aboard the *Centurion* for provisions and then put ashore, landed it and then took off the men. Our people set fire to the shipping and then took to the woods. Our company [Capt. Peter Warren's] encamped in the woods. Took what provisions we could carry. Had 4 prisoners to guard.

Sunday, Aug. 15. Took up our line of March at daybreak, lost our way and came across about 200 of our Regt. And sailors and marines. Went across a large meadow; struck a road in the woods and kept on till 7 o'clock; took breakfast and proceeded on to Belfast where we arrived at 12 o'clock. Exceedingly warm. Came to a river and crossed in canoes. Capt. Warren purchased 2 sheep and paid 18 dollars for them. Took dinner. Arrived at a fine plantation and had a good dish of tea. Gen. [Peleg] Wadsworth and Capt. [Ebenezer] Buck supped with us. Had a fine barn to sleep in and rested comfortably.

Aug. 16. Marched early through marshes, beaches and thick woods, over mountains and valleys to Ducktrap [Northport] where we arrived, the sun an hour high. P.M. One of our prisoners deserted this morning.

Aug. 17. Set off early and traveled by the shore. Halted by Gen. Wadsworth's orders. Arrived at the westerly part of Camden at 1 o'clock. The place called Clam Cove. [Went to] Headquarters and drew an allowance of fresh beef. Turned out a Sergeant's Guard and took possession of a large barn for our barracks.

Aug. 18. Heard that Gen. Lovell and Admiral Saltonstall were taken by the enemy. [A rumor only.] Capt. [William] Cobb and his company arrived here at 12 o'clock. [Daniel] Mussey started for Falmouth.

Aug. 19. Mr. [Somers] Shattuck and Stephen Tukey arrived this morning, says Woodbury Storer was taken on board the Hampden. Mr. Shattuck and Houchin Tukey started for home. Order for Capt. Warren to march to West Shore South West Gigg. [Stephen Tukey was the son of John and Abigail (Sweetser) Tukey of Falmouth Neck, and was born July 6, 1754, married, in 1780, Hannah Cushing, and died July 8, 1826. He was the writer's great grandfather. Houchin Tukey was his brother.]

Aug. 20. Marched to Col. [Mason] Wheaton's, 6 miles. Set a corporal's guard. Here is a double saw mill and grist mill.

Sunday, Aug. 22. Lieut. [Peter] Babb set off for home or Falmouth with some four men because we had no provisions. [Zach.] Baker, [John] Clough, Thomas Harper, [Benjamin] Mussey and myself [William Moody] started for St. George between 11 and 12 o'clock.

Aug. 24. Arrived at New Meadows and put up at one Capt. Curtis' where we were hospitably entertained.

Aug. 26. Capt. Warren arrived home, [and probably the whole company].

PROCLAMATION

By SOLOMON LOVELL, Esq., Brigadier-General and Commander in Chief of the Forces of the State of Massachusetts Bay, and employed on an Expedition against the Army of the King of Great Britain at Penobscot. July 29, 1779

WHEREAS it hath been represented to Government that an armament of some sea and land forces belonging to the King of Great Britain, under the encouragement of divers of inhabitants of these parts, inimicably disposed to the United States of America, have made a descent on Penobscot, and the parts adjacent; and after propagating various false reports of a general insurrection of the Eastern and Northern Indians in their favour, a Proclamation has been issued on the 15th of June last, signed Francis McLean and Andrew Barclay, said to be in behalf and by authority of said King, promising grants of lands which he never owned, and of which he has now forfeited the jurisdiction by an avowed breach of that compact between him and his subjects, whereon said jurisdiction was founded, and terrifying by threatenings which his power in this land is unable to execute, unless his servants have recourse to their wonted methods of midnight slaughter and savage devastation, all designs to induce the free inhabitants of these parts of the State to submit to their power, and to take an oath of allegiance to their King, whereby they must greatly profane the name of God and solemnly entangle themselves in an obligation to give up their cattle, provisions and labour to the will of every officer pretending the authority of said King, and finally to take arms against their brethren when ever called upon; and it appears some persons have been induced out of fear and by the force of compulsion, to take said oath, who may so far be imposed on as to think themselves bound to act in conformity thereto :

I have thought proper to issue this Proclamation, hereby declaring that the allegiance due to the ancient constitution obliges to resist to the last extremity the present system of tyranny in the British Government, which has now overset it ; that by this mode of government the people have been reduced to a state of nature, and it is utterly unlawful to require any obedience to their forfeited authority; and all acts recognizing such authority are sinful in their nature; no oaths promising it can be lawful; since if any act be sin in itself, no oath can make it a duty; the very taking of such an oath is a crime, of which every act adhering to it is a repetition with dreadful aggravations.

In all cases where oaths are imposed, and persons compelled to submit to them by threats of immediate destruction which they cannot otherwise avoid, it is manifest that, however obligatory they may be to the conscience of the compeller, whose interest and meaning is thereby so solemnly witnessed, it can have no force on the compelled, whose interest was known, by the compulsion itself, to be the very reverse of the words in which it is expressed.

At the same time, I do assure the inhabitants of Penobscot and the country adjacent, that if they are found to be so lost to all the virtues of good citizens as to comply with advice of said pretended Proclamation by becoming the first to desert the cause of freedom of virtue and of God, which the whole force of Britain and all its auxiliaries now find themselves unable to overthrow, they must expect also to be the first to experience the just resentment of this injured and betrayed Country, in the condign punishment which their treason deserves. From this punishment their invaders will be very unlike to protect them, as it is now known they are not able to protect themselves in any part of America. And as the protection on which those proclaiming Gentlemen say they have power only to promise, can be afforded by nothing but the forces which they command, and of these forces by the blessing of God, I doubt not in a very short time to be put in possession; so there is more reason to expect it from the Indian members of the community and treated accordingly, anything nations around, as good part of them are now in my encampment, and several hundreds more on their way speedily to join me; and I have the best evidences from all the rest, that they steadfastly refused to accept of any presents, sign the papers, or do any the barbarous acts assigned them by our Enemies; and on the contrary hold themselves in readiness, on the shortest notice, to turn out for the defence of any place which these men may attack.

Therefore, as the authority committed to me necessitates my executing my best endeavours to rid this much⁻ abused country, not only of its foreign but also from its domestic enemies, I do, therefore, declare that when, by the blessing of Heaven on the American arms, we shall have brought the forces that have invaded us to the state they deserve, it shall be my care that the laws of this state be duly executed upon such inhabitants thereof as have traitoriously abetted or encouraged them in their lawless attempts.

And, that proper discrimination may be made between them and the faithful and liege subjects of the United States, I further declare that all persons within the Eastern country, that have taken the oath prescribed by the Enemy, and shall not within forty-eight hours after receiving notice of this proclamation repair to my camp at Majabigwaduce, with such arms and accoutrements as they now possess, shall be considered as traitors who have voluntarily combined with the Common Enemy in the common ruin; but all such as shall appear at headquarters within said term, and give proper testimony of their determination to continue cordially in allegiance to the United States of America, shall be recognized as good and faithful members of the community and treated accordingly, anything obnoxious in their taking the oath notwithstanding.

Given at the Head-Quarters on the Heights of Majabigwaduce, this 29th day of July, Anno Domini, 1779, and in the Fourth Year of the Independence of America.

(Signed) S. LOVELL,

Brig. Gen.

By Command of the General,

JOHN MARSTON, Secretary.

Letter – August 11, 1779

Copy of General LOVELI/S Letter to
Commodore SALTONSTALL; taken with other
Papers on board the Transport.

Head Quarters, Majabigwaduce Heights, Aug. 11, 1779

SIR,

In this alarming posture of affairs, I am once more obliged to request the most speedy service in your department; and that a moment be no longer delayed to put in execution what I have been given to understand was the determination of your last council. The destruction of the Enemy s ships must be effected at any rate, although it might cost us half our own; but I cannot possibly conceive that danger, or that the attempt will miscarry. I mean not to determine on your mode of attack; but i appears to me so very practicable that any further delay must be infamous; and I have it this moment by a deserter from one of their ships, that the moment you enter the harbour they will destroy them; which will effectually answer our purpose.

The idea of more batteries against them was sufficiently reprobated; and, would the situation of ground admit of such proceeding, it would now take up dangerous time; and we have already experienced their obstinacy in that respect.

You cannot but be sensible of my ardent desire to co-operate with you; and of this the guard at Westcot' s is a sufficient proof, and which I think a hazardous distance from my encampment. My situation is confined; and while the Enemy s ships are safe, the operations of the Army cannot possibly be extended an inch beyond the present limits; the alternative now remains, to destroy the ships, or raise

185

the siege.

The information of the British ships at the Hook* *(probably sailed before this) is not to be despised; not a moment is to be lost; we must determine instantly, or it may be productive of disgrace, loss of ships and men; as to the troops, their retreat is secure, though I would die to save the necessity of it.*

I feel for the honour of America, in an expedition which a nobler exertion had long before this crowned with success; and I have now only to repeat the absolute necessity of undertaking the destruction of the ships, or quitting the place; and with these opinions I shall impatiently wait your answer.

> *I am,*
>
> > *Sir,*
> >
> > > *Yours, etc.,*

> > > > *S. LOVELL, Brig. Gen.*

* Sandy Hook, New York

Journal Chronicling Col Mitchell's Regiment's Involvement

July 16, Col. Mitchell's men were embarking on the transports awaiting at Falmouth Neck, now Portland, and by the nineteenth were ready for departure, when Capt Abner Lowell fired from the battery, a gun as a signal for the fleet to set sail for its destination.

The transport on which Capt. Peter Warren's Falmouth company was quartered was the sloop Centurion, eighty and one-third tons, Capt. William McLellan of Falmouth Neck. She carried three men for a crew beside the captain. Among the stores sent on board for the use of the crew were seven gallons of rum. This sloop was destroyed with the others, and was appraised at twenty-nine hundred pounds for the settlement with her owners.

July 19, Col. Mitchell's regiment arrived at Townsend Harbor now Boothbay Harbor, the rendezvous of the expedition, in the evening of July 19, having sailed from Casco Bay in the morning. Gen. Lovell made his headquarters at Rev. Mr. Murray's house, where the returns of the regiments were examined by him on the twenty-first. The next day the troops were reviewed by the commander-in-chief, which must have been unsatisfactory to him as the men had had no opportunity to learn discipline, and but few had any knowledge of the manual of arms. The twenty-third there was an unfavorable wind and the expedition remained in the harbor, waiting for a favorable opportunity to sail.

July 24, the whole expedition set sail for Penobscot Bay. The fleet made an imposing appearance as it sailed out of Boothbay Harbor along the coast into the Penobscot. The men on board were in high hopes of success. The fleet came to anchor under upper Fox Island that night. Here they were joined by a party of Penobscot Indians, who reported that Gen. McLean, the British commander, had tried to tamper with them, but to their honor it can be said that they remained true to their promise made in 1775. Our commanders soon learned that the British were entrenched at Bagaduce, and had three sloops of war in command of Capt. Henry Mowat, so well remembered in the history of Portland.

July 25, The next day, found the Americans in range of the guns of the enemy, who commenced firing from the shore, whereupon our armed vessels fired several broadsides at their forts. The British, however, prevented the landing of our boats that night, but the next day, July 26, the vessels warped in, and about noon an attempt was made to land. This was also unsuccessful. About six o'clock that afternoon, while putting off from Nautilus Island where the Americans had made a landing, a boat was struck and Major Daniel Littlefield of the York detachment and two men were drowned.

July 28, about two hundred of the militia and a little over that number of marines were ordered to land under the guns of the fleet, and the movement was begun about half an hour before sunrise, when the British in ambush opened a galling fire upon the Americans, killing several, among whom was

Major Welch of the marines.

Our troops replied with effect. A landing was made on the shore under the steep bluff now called "Trask's Rock," at Castine, on the western side of the point. This bluff is one hundred and fifty to two hundred feet high or more at some points. Castine Past and Present says:-- "Where the marines made their ascent was quite precipitous for some thirty or forty feet and after that the ground is still rising for some distance and was covered with boulders." The marines and militia divided themselves into three parties, when a most gallant assault, without order or discipline, each man dependent on his personal courage, was made on the enemy above, against a most destructive fire, which they were in no position to return. In twenty minutes our troops were at the top occupying the British ground. The first company to reach that point was Capt. Peter Warren's Falmouth company. During this time our fleet was bombarding the enemy's forces. Gen. Lovell wrote in his journal:--

When I returned to the shore it struck me with admiration to see what a precipice we had ascended, not being able to take so scrutinous a view of it in time of battle; it is at least where we landed three hundred feet high and almost perpendicular & the men were obliged to pull themselves [up] by twigs and trees. I don't think such a landing has been made since Wolfe.

Our loss in the assault is variously stated, Gen. Lovell gives fourteen killed and twenty wounded, while Gen. Wadsworth says it was about one hundred, which is repeated in most of our histories. The marines suffered the most. It has been truly said that this was the bright spot in the expedition and that no more brilliant exploit than this was accomplished by our forces during the whole war." It was a trying ordeal to the undisciplined and untried militia and marines, but they exhibited the resolute courage of the American soldier. If the whole expedition had been successful, our histories would have resounded the praises of Gen. Lovell and his men.

Soon after the Americans reached the top of the bluffs they threw up entrenchments so that they might be able to hold the ground they had so heroically gained.

Immediately after this engagement a council of war of the American land and naval forces was held. The officers of the army were in favor of demanding the immediate surrender, but Commodore Saltonstall and some of his officers were opposed to it. Then the army was for storming Fort George, but because the marines had suffered so much in the assault, the com-modore refused to land any more and even threatened to recall those already on shore. Then it was that it was decided to send to Boston for reenforcements, which resulted in the starting of Col. Henry Jackson's Continental regiment for their relief.

July 29, the enemy shelled the Arnericans who maintained their position. In the early morning of the thirty-first a party of soldiers under Gen. Wadsworth captured a redoubt with but a small loss, taking several prisoners. It was here that Major Samuel Sawyer, sometimes written Sayer, was mortally wounded. He was "a brave and worthy officer," and belonged in Wells, Maine.

August 2, Rev. John Murray, of Boothbay, then Townsend, who had been induced to join the expedition as the chaplain of Col. McCobb's regiment, volunteered to carry despatches from the general to the government at Boston. August 6, Major David Brandish, of Falmouth Neck, also started for Boston with despatches. He was on a visit to the army at the time and not connected with it. The next day, two men of Col. McCobb's regiment were punished for desertion by riding a wooden horse twenty minutes, with a musket attached to each foot.

There was more or less fighting along the line until the seventh, but no general assault was made. On that day a detachment of Americans advanced against the enemy's position to draw them out, but without success. August 9, an attempt was made by our soldiers to land on Hyannis Point, but with no success. August 13, an effort was made to bring on an engagement with the British, which was also unsuccessful. Then it was that our troops actually took the rear of Fort George, but did not get possession. The delay gave the British every advantage.

August 3,The next day came the startling news that a British fleet was at the mouth of Penobscot Bay with reenforcements. Upon the approach of the fleet, Commodore Saltonstall formed his vessels across the bay in the form of a crescent, to check their advance sufficiently to allow the land forces time to make their escape. The British commander, Sir George Collier, feeling such entire confidence in the superiority of his fleet, advanced without hesitation and poured a broadside into our vessels, whereupon they crowded on all sail in an attempt at an indiscriminate flight. The Hunter and Hampden were taken and the balance of the fleet was burned or blown up by their crews.

Gen. Lovell in his journal said :--

The Transports then again weigh'd Anchor, and to our Great Mortification were soon follow'd by our fleet of Men of War persued by only four of the Enemv's Ships, the Ships of War passed the Transports many of which got a Ground & the British Ships coming up the Soldiers were obliged to take to the Shore & set fire to their Vessells, to attempt to give a description of this terrible Day is out of my Power it would be a fit Subject for some masterly hand to describe it in its true colours, to see four Ships persuing seventeen Sail of Armed Vessells nine of which were stout Ships, Transports on fire, Men of war blowing up, Provision of all kinds & every kind of Stores on Shore (at least in small Quantities) throwing about, and as much confusion as can possibly be conceived.

The destruction of the vessels engaged in this expedition was the end of Massachusett's separate naval force and reduced the national navy of the United States to the very lowest terms. Our commodore had stubbornly refused to cooperate with the land forces at the proper time and the result was a terrible disaster to the Americans. The army, with the men of the fleet, retreated up the river with little order, Each one looked out for himself and his own safety. They fled to the woods and carried scanty provisions which lasted but a few days, when the men were obliged to subsist on whatever they could find on the way, until they reached their homes. Some fell by the wayside and perished from starvation and exposure, and many who returned home filled premature graves as the result of the hardships they were obliged to endure. Many of the men said little about their sad experience in this retreat, because it revived unpleasant mernories of a service which was a great disappointment to them and for which they were in no way to blame3.

Standing on this battlefield a few year since, after making the ascent of the bluff where the brilliant assault was made, and while looking out over the surrounding country and bay, I was carried back in my mind to the summer of 1779 when the events that made that ground historic occurred. Not one was then living to tell the tale. An aged uncle had told me that when a little boy his grandfather had said to him that he was a soldier in that expedition. I was now at the place where my grandmother's father, Stephen Tukey, had fought to drive the British from our state of Maine. There were hundreds of grandfathers there and in justice to their patriotism and loyalty to their country, this history is written that their names may not he forgotten, but be preserved to receive the reverence of their descendants and a grateful people.

The following returns of the army at Bagaduce have been preserved:--

July 20, 873 men fit for duty.

July 31, 847 men fit for duty.

Aug. 4, 762 men fit for duty.

Aug 7, 762 men fit for duty.

Another return was made with no date which gave nine hundred and twenty-three men fit for duty, But at least two more companies had joined the army and probably more.

Williamson says of the retreat:--

> Guided by Indians they proceeded in detached parties suffering every privation. For, not being aware of the journey and fatigue they had to encounter, they had taken with them provisions altogether insufficient, and some who were infirm or feeble actually perished in the woods. A moose, or other animal, was occasionally killed which being roasted upon coals was the most precious, if not the only morsel, many of them tasted during the latter half of their travels.

The Bagaduce expedition was such a subject of obloquy and remark that the General Court appointed a committee of investigation into the cause of its failure. Gen. Artemus Ward was the president of that committee. Col. Jonathan Mitchell of this regiment stated before them that it was not in the power of Gen. Lovell at any time, with his army, to have reduced the enemy while they were on the ground. He also said that if the British shipping had been destroyed and the land forces had been aided by men from the fleet, armed with muskets, they could have destroyed the enemy. He thought that the British fleet could have been crushed any day before they were reenforced.

In regard to the retreat Col. Mitchell said:-

> About one o'clock in the morning of the 14th, I went to Gen. Lovell's marquee. He ordered me to get my regimental baggage and camp

*equipage to the shore and have my men ready
for marching. I did so and at break of day was
ordered to march, and at about sunrise
embarked them on board transports and
proceeded up river above the old forts. There
received orders from Brig. Wadsworth through
the adjt gen. to repair to a certain height, there
to receive and retain as many of the army as
came that way. I repaired to the spot myself
but I found no men there but the matrossis
and Capt. Cushing with them, from the time of
our retreat to this time. I had not issued any
orders to my regiment not to disperse or to
repair to any particular place but only to go up
river. I tarried till about sunset and no men
came that way, then I went into the woods to
look up my regiment. On the sixteenth about
nine in the morning I set off for home, but
without leave from any superior officer. The
eighteenth I arrived at the Kennebec River; on
the twentieth, at night I reached home and on
the twenty-first, went to Fort Weston (Augusta)
without any men.*

Adjt. Gen'l Hill said of the soldiers of the
expedition:

*"If they belonged to the train hand or alarm list
they were soldiers, whether they could carry a
gun, walk a mile without crutches or only
compos mentis sufficient to keep themselves
out of fire and water." The "soldiers were very
poorly equipped, the chief of them had arms
but many of them were out of repair and very
little or no ammunition. Most of the officers, as
well as the men were quite unacquainted with
any military maneuvers and even the manual
exercises."*

The night before the assault of July 28, the
soldiers had no sleep. Adjt. Hill said that "Col.

Mitchell's officers were so terrified at the idea of storming that they found fault with the colonel's nominations and absolutely drew lots on the parade who should go to take command of the men and included those then on guard, and relieve them if it fell to any of their turns." He also stated that " the troops behaved with spirit as far as came to my knowledge, but without any order or regularity find it was with great difficulty that we got them into any order or form of defense after we got to the heights."

August 10, Gen. Lovell called for six hundred volunteers, to test the temper of the troops. Col. Mitchell's regiment was the only one that filled its quota, which was two hundred. Adjt. Hill said that "Col. Mitchell got his 200 with great difficulty, including boys, old men and invalids." The men got the impression that these volunteers were for a general assault on the British works, the principal of which was Port George. The other regiments had even more difficulty, At the same time, Col. McCobb could get but one hundred and forty-six volunteers from his Lincoln County regiment, and Major Cousins had twenty men desert from his York County battalion, and it took so many men to pursue them that he could not furnish his quota. Of a muster of six hundred ordered, only four hundred were secured, which was all the officers said they could find fit for duty. One half of these were from Mitchell's regiment.

The committee after hearing the testimony of the general and regimental officers, and the commanders of the armed vessels, pronounced as their opinion, that *"the principal reason of the failure was the want of proper spirit on the part of the commodore."* He was blamed for not *"exerting himself at all in the time of the retreat by opposing*

196

the enemy's foremost ships in pursuit." They also stated *"that Gen. Lovell throughout the expedition and retreat acted with proper courage and spirit, and had he been furnished with all the men ordered for the service or been properly supported by the commodore he would have probably reduced the enemy."* Also *" that the naval commanders each and every one of them behaved like brave experienced officers during the whole time."* Then they said that *"Brigadier Wadsworth, the second in command throughout the expedition, in the retreat and after, till ordered to return to Boston, conducted with great activity, courage, coolness and prudence."*

After hearing the whole report, from which the above are but quotations, the General Court adjudged *"that Commodore Saltonstall be incompetent ever after, to hold a commission in the service of the state and that Generals Lovell and Wadsworth be honorably acquitted."*

Letter of General Wadsworth Regarding the Expedition

Gen. Wadsworth in a letter to William D. Williamson, dated January 1, 1828, said of the Penobscot Expedition of 1779:--

In the first place the want of a sufficient land force was a probable cause of the failure. We had less than 1000 men, where 1500 were ordered by the State authority; whose fault this was I know not; but so it was. This was just about the Number of the Enemy; but they were disciplined Troops & fortified with a simple redoubt, which was good however against n simple assault. Our Troops were entirely undisciplined, having never been paraded but once, on their passage down, being put in to a harbour by head Wind; I think at Townsend, nor had these Men ever had the chance for discipline that our western Militia had; however they were generally brave &spirited Men. Each in his own opinion willing to encounter two of the Enemy, could he have met them in the bush; and would our numbers have justified an Attack, I have no doubt that they would have given the Enemy a brave Assault. Although our numbers were small our Fleet had an imposing appearance, I think the Enemy must have reconed upon at least 3,000 men from the appearance of our Transports.

The same Morning of our Landing a Council was called of officers, both land & naval. Some of the land officers were for summoning the fort, giving them honorable Terms, whilst others disuaded from the Measure alledging that in case of a non complyance We should be in a bad predicament; the Commodore and the naval Officers were generally against the Measure; as his officers were chiefly commanders of Privateers bound on a Cruize as soon as the siege was over. The Commodore also refused to lend any more of his Marines in case of Assault and was about to recall the 200 marines which he had lent on our first landing. They had suffered great Loss in the landing. This seemed to put the Question of Storming the Fort out of the Question. The next Question was, what then shall be done? & it was concluded to send off two Whale Boats to the Gov'r & Council with the intelligence of our situation and request a reinforcement while we kept our possession in the face of the Enemy & trust to the event of a reinforcement to the Enemy & of ourselves. In the meantime we reduced our out Posts & Batteries, destroyed a considerable Quantity of Guns, spiked their cannon in all their out works & gave them fair opportunity of Sallying if they chose it.

In the meantime we were employed daily, or rather Nightly in advancing upon their Fort by Zigzag intrenchments till within a fair gunshot of their Fort so that a man seldom shew his Head above their Works. Whilst thus lying upon our Arms it was urged upon Genl Lovell to erect some Place of resort up the river at the Narrows, in Case of Retreat so that the Troops might have a place of resort in case of necessity & also to have some place of Opposition to the Enemy should He push us thus far – but the Genl would hear nothing of the kind; alledging that it would dishearten our Army & shew them that we did not expect to succeed-- & forgetting the good old Maxim "to keep open a good Retreat."

Had the Genl and Commodore kept upon a good understanding with each other & had they co-operated with each other they would have probably stormed and carried the Enemy's Post; & been off before there was any danger of the arrival of the enemy's reinforcements. Here we may see the policy of securing a place of Retreat. The fleet might have been saved, the Army kept together and marched in *a body wherever wanted, instead of scattering, starving, &c.*

Here we had been laying upon our Arms almost inactive 14 days when our Spy Vessels bro't the news of a Large Fleet approaching, which might be expected the next day, if the south wind should prevail. Genl. Lovell was now on board the Warren, Commodore's Frigate and sent his Orders to me to retreat with all possible dispatch, which was effected without leaving a canon or a pick axe behind, the Enemy's Fleet in full view standing up with full sail & much superior to ours in Appearance. As soon us the Troops, the Cannon and all our implements of War, with the Hospital, were on board, the Transports stood up the River -- O, then how we wished for a place of rendezvous, the Transports might have been saved. Our Fleet soon persued the Course of the transports, but soon went theirs, forcing their way through the Narrows against a strong tide with Oars & Studen sails all sett, wllilst part of our Transports had run on shore just at the foot of the Narrows. The troops landed, the flames bursting forth from the midst of them, set by their own Crews. The Enemy persuing to within Cannon Shot, but unable to persue farther against a strong tide, left those that would be persuaded to enter the Transports & rescue a small Quantity of provisions for the retreat & to collect and embody themselves for their own safety. Three or four Companies were thus kept together with which I marched the next morning for Camden, where they arrived the second day & made a stand. The rest of the Troops went up the River in the Vessels of War & Transports landing as they saw fit & then Genl Lovell under the guidance and Assistance of the Indians made his way from the head of the Tide in the Penobscot over to the Kennebec; & in about a fortnite arrived at Towrnsend when was the first that I had seen or heard front him since ordering the Retreat. That part

200

of the Fleet that got up the River ahead of the Enemy were either burnt or destroyed by their own crews making their way thro the woods for the Kennebec in a starving condition. Had Genl Lovell been furnished with the Number of Militia which was at first proposeh, or had He been appointed to sole command of both Army & Navy, I think it highly probable that he would have reduced the Enemy for He was a Man of Courage & proper Spirit, a true Roman Character, who never would flinch from Danger; but He had not been accustomed to the Command of an Expedition in actual service. The Commodore did not feel himself so much engaged in the Cause.

Not that he was, in my opinion, a Coward, but willful & unaccommodating, having an unyeilding will of his own. -- Genl Lovell was n very personable Man, I should judge about 50, of good repute in the Militia, as well as Senate, a Farmerer by profession & I believe lived in Weymouth. Commodore Saltonstall about the same age, of New Haven, Ct. Report said that he fought a very good battle afterward in a large Privateer whicll shew him to be a Man of Courage. The command of a Fleet did not set easy upon his shoulders tho he could fight a very good Battle in a single Ship.

Here it may be not improper to mention that the Action at our landing on Bagaduce might have been called brilliant, had the event of the Enterprise been fortunate. But let military men not talk of glory who lack success. It was on the dawning of the third day after our arrival (the second was prevented by the surf occasioned by a brisk south wind). The morning was quite still but somewhat Fogy. The Vessels of War were drawn up in a Line just out of reach of Musket Shot & 400 Men (viz. 200 of Militia & 200 Marines) were in Boats along side ready to push for the Shore on Signals. The highest Clift was prefered by the commander of the Party, knowing that his men would make the best shift in rough ground. The fire of the Enemy opened upon us from the top of the Bank or Clift, just as the boats reached the Shore. We step'd out & the boats immediately sent back. There was now a stream of fire over our heads from the Fleet & a shower of Musketry in our faces from the Top of the Cliff. We soon found the Clift unsurmountable even without Opponents. The party therefore, was divided into three parts, one sent to the right, another to the left till they should find the Clift practicable & the Center keeping up their fire to amuse the Enemy. Both parties succeeded & gained the Height, but closing in upon the Enemy in the Rear rather too soon gave them opportunity to escape, which they did, leaving 30 kill'd, wounded & prisoners. The conflict was short, but sharp, for we left 100, out of 400, on the shore & bank. The marines suffer'd most, by forcing their way up a foot Path leading up the Clift. This Action lasted but 20 nIinutes &- would have been highly spoken of, had success finally crowned our Enterprise.

The valuable letter, from which the above is quoted, was written to Mr. Williamson while he was preparing his history of Maine, which was published in 1832. Gen. Wadsworth was then nearly eighty years of age, and the events happened over forty-eight years before. The letter was contributed by Dr. John S. H. Fogg and published in the Maine Historical Society Collections, Vol. II, Series II, Fol. 153.

Gen. Peleg Wadsworth was a member of Congress fourteen years, 1792-1806, and retired at his own request. He built the first brick house in Portland, in 1785 and 1786, then of but two stories, now known as "Longfellow's Home." He removed to Hiram, Maine, in 1806, where he died in 1829, aged eighty-one years. His sons, Henry and Alexander Scammell Wadsworth, were gallant officers of the American navy.

Col. Enoch Freeman's letter to the Council at Boston

Col. Enoch Freeman sent the following letter August 18.

For the first five years of the Revolutionary war Massachusetts was governed by a committee of the Council.

FALMOUTH, Aug. 18, 1779.

SIR:--The invasion of the Penobscot under a very considerable force of the enemy, their progress there and the ravages committed by them in other places at the Eastern part of this State make us apprehensive that they have a design to cut it off from the other part of the State and either annex it to the province of Nova Scotia, or form it into a separate government under the British Administration.

Under such apprehensions, a number of gentlemen from most of the towns in this County, this day assembled in Convention in this town, to consult what is proper to be done for our safety and defence.

We think that the Harbor here would be of such importance to the enemy, in the execution of what we judge to be their grand design, that they will not much longer neglect to attempt to possess themselves of it, and make it a place of Rendezvous for their troops and ships of Force. -- And we are sorry to inform your Honors that such is the state of our fortifications and such the weakness of our Force, that unless some measures are immediately entered into for our protection and defence, we fear we shall fall a prey to their rage and malice. We therefore humbly pray that your Honors would take our case into your serious consideration and order that such steps may be taken as will put us in a good position of defence.

We have recommended to the several towns in this County to raise immediately their respective proportions of one hundred men, to repair the forts here and build others in such places as a Committee (whom we have appointed for the purpose) shall best judge, and we trust the General Court will make provisions for paying them for their services.

And we would request that the Honorable Council would appoint and send as soon as possible, some experienced faithful engineer to take the oversight of the work.

We would further pray that at least two hundred men might be ordered here from the County of York or some other County to the southward of us, to increase our strength, which is already much reduced.

We also think it necessary that a number of cannon and a suitable quantity of military stores should be procured and sent here to be placed in such Forts as may be erected, and also field pieces, (two we think necessary) And as provisions are extremely scarce here and it would be almost impossible to collect on an emergency as much as might be wanted, we think it absolutely necessary that a Magazine thereof should be provided and placed in a proper part of the town, to be used when an alarm should require it.

I am, in the name and behalf of the Committee, your Honor's most obedient and, humble serv't

ENOCH FREEMAN.

Rev. John Murray's Letter

The Rev. John Murray, the chaplain, wrote from Brunswick to Jeremiah Powell, Esq., under date of August 21, 1779:--

Our case is very bad. Hundred of families are now starving in the woods, their all left behind them, all will despair and the majority will quit the country and the rest will revolt if something vigorous be not done to protect them from the insolence of the triumping foe who are carrying fire and desolation wherever they come. A large reinforcement of men, intrenching tools, cannon, ammunition and provisions is absolutely necessary to save us. Not a moment is to be lost. A very little delay will put us beyond remedy, but if we are immediately relieved this little disaster need not discourage us. It will, if we act with proper spirit, issue in our good.

The original letter is in the Massachusetts Archives, Vol. CXLV, Page 140.

In the latter part of August, Col. Mitchell's regiment had reached Falmouth Neck, but arrived there in a disorganized and demoralized condition. Three companies were retained for a garrison and twenty men of Capt. Curtis company were stationed at Harpswell. All others were discharged.

Col. Henry Jackson's Letter upon arriving at Falmouth August 1779

Col. Henry Jackson's Continental regiment, which had been ordered from Rhode Island to reenforce the expedition, learned of the disaster off Kittery while on their way. They went into camp at that place. It was then thought that the British, elated at their success, would proceed to Casco Bay and attempt the capture of Falmouth Neck.

This of course caused much alarm at that settlement. Col. Jackson's regiment was ordered to march to Falmouth, from Kittery, and arrived there the twenty-seventh, and went into camp on Munjoy Hill, above the Eastern Cemetery, much to the relief of the inhabitants.

This regiment had four hundred well uniformed and equipped men and had then participated in the battles of Monmouth and Quaker Hill. This was probably the only fully uniformed and equipped regiment the people of Falmouth saw during the war.

Col. Jackson wrote the next day after his arrival:--

I find this town and harbor is by no means in a state of defence as but a few of the cannons are fit for any long service. To make this post defencible it will be necessary to have a number of heavy cannon immediately sent here: the militia are exceedingly destitute of arms, ammunition and accoutrements as I find by euquiry that not more than one-half are armed or accoutered.

The committee of safety of Falmouth addressed the following letter to the Council:--

Falmouth, 30 August, 1779.

Sir: The Committee of Safety &c for Falmouth would

inform the Honorable Board of their embarresments and beg their direction. The return of the seamen from Penobscot in the greatest distress imaginable has obliged us to act as commissary, quartermaster, &c, &c. To furnish them with necessary provisions and to relieve their distresses we have been obliged to issue some impress warrants: some provisions we have purchased and some we have borrowed. We have observed the strictest economy and order that necessary confusion would admit of; the men returned without officers, without orders.

We shall transmit an account of our doings as soon as the men have been done returning.

Col. Jackson applies to us for some assistance where he has not proper officers to supply them. This however gives us but little trouble: but the militia who have returned from Penobscot are ordered to this place: they are not properly attended with their officers and those who do attend them have not proper directions what to do with their men: they apply to the Committee. The Committee know of no business they have with them; here we are much embarrased. We have also frequent applications from expresses for assistance, or sometimes are obliged to send off expresses ourselves New applications of various kinds, are daily made to us and new difficulties arise. In short, affairs here are in the wildest confusion. We wish for the direction and assistance of the Hon. Board.

We are &c
The Committee of Safety &c for Falmouth,
Stephen Hall, Chairman.
Hon. Jer. Powell,
Pres. of Council.

Col. Jackson's regiment started on their march to Boston, September 7, as all danger of an attack seemed to be over; but a portion of Col. Mitchell's still remained.

Falmouth Selectman's Letter to the Council in Boston

The selectmen and the committee of the town sent the following letter to the Council:--

FALMOUTH, SEPT. 13, 1779.

To the honorable Council of State of Massachusetts Bay.

The Selectmen and Committee of Safety at Falmouth beg leave to inform the Honorable Board of the receipt of their letter of the 3rd inst. The enclosed directed to Brig'r Thompson was immediately forwarded. It is now seven days since, but we have not heard of his taking any measures towards raising the three hundred men to be stationed at Falmouth.

The letter from the Hon. Council to Col. Jackson in his absence we took the liberty to open: in answer to which we would inform your Honors that the Regt. From Penobscot was ordered by Gen'l Lovell to Falmouth to guard from this place to Harpswell and to be under the direction of the Committee of Safety at Falmouth.

As a greater part of these were destitute of arms and accoutrements, the Committee thought proper to discharge five of the companies except 20 of the company commanded by Capt. Curtis of Harpswell, who are kept guard at that post. The remaining three companies which were best armed and accouted are now stationed at this place and at Cape Elizabeth.

We esteem it a duty incumbent on us to inform your Honors that the militia in this County are at present in a situation incapable of defending us in the case of an attack, principally owing to their ignorance and neglect of some of the principal officers of the Brigade.

A convention of this County is to be held next Friday when proper representation of the state of the militia will be made to the Hon'ble Court. We are with sentiments of respect

Your Honors most obed't serv't,

By order in behalf of the Selectmen,

Benjamin Titcomb.
And Committee of Safety
Stephen Hall Chariman.

P.S. A number of small arms and cartridges has lately been received, also a quantity of ordinance goods, addressed to Col. Jackson by the Board of War a particular return of which shall be made by the first opportunity.

In Council, Sept. 22, 1779. Read and sent down.

John Avery D. Secr'y.

Massachusetts' General Court Conducts Hearings

Results of the General Court's Investigation

The General Court's investigating committee concluded that if Brigadier General Lovell had the full complement of 1,500 militia he would have had success. Yet the fact that Maine could raise such a force was a large assumption by Massachusetts.

As a point of reference, on 9 June 1779, Massachusetts issued quotas to the counties in order to fill fifteen Massachusetts battalions for service in the Continental Army. Of the two thousand men levied, the quota for the three Maine counties equaled two hundred and forty-seven men.

The requirements for the Penobscot Expedition showed a six-fold increase over the anticipated soldier requirements earlier that month. Locally, Lincoln County, which encompassed the Penobscot River, saw a nine-fold increase in its quota of men. Given the emergency conditions, it may seem sensible that the people most affected by the British invasion would furnish the most men.

It is difficult to assess Maine's ongoing contribution to the war in 1779 because its soldiers served in Massachusetts's regiments and not in specific Maine regiments. The Maine Historical Society estimates Maine conducted six thousand enlistments throughout the war, not including reenlistments. Of those, the society confirmed nearly eleven hundred of them served at Valley Forge during the winter of 1777 and 1778.

In 1779, of the 247 men (by county is York-101 men, Cumberland-80, and Lincoln-66) from Maine served as part of Massachusetts' regiments in Rhode Island, in George Washington's Army in New

York, and as part of General John Sullivan's Expedition.

Additionally, local defense demanded more troops from the militia manpower pool. The town of Machais, precariously positioned on Maine's ill-defined border with Nova Scotia and Falmouth, both objects of previous British attacks, maintained their own active garrisons of militia. Other towns maintained their own small forces as protection against British foraging parties or even worse, raids conducted by American privateers.

The account of Colonel John Brewer (then a captain), founder of the present day city of Bangor, demonstrates the split loyalties of the militia. As a commander of a militia company, he and his men did not answer the call to serve in the Penobscot Expedition. Instead, the town committee decided to weigh its own options, sending him to meet with General McLean personally.

He even conducted a second visit to McLean the day prior to the American's arrival. Afterwards, he hailed the ships of the expedition and met with his brother, Colonel Josiah Brewer. His brother took him to Lovell and Saltonstall in order to give them intelligence on the British defenses.

Completing that, he returned home with orders from his brother to return with half of his militia company. His response demonstrates his split loyalties, which no doubt mirrored that of the local population in general. *"This order I obeyed; but my family not then being in a situation to leave, my men were put under the command of another captain, and I returned home for one week, when I again repaired to my post."*

Later, after he returned and the siege wore on, he noted "nothing important appearing to be going on, [so] I again returned home."

Living just miles up the Penobscot River, the British fort was certainly a direct threat, yet both Brewer and his community remained cautiously neutral.

While Massachusetts's militia laws clearly provided for the impressment of men into military service, it was far from the preferred method of doing so. Massachusetts had long relied on enlistment bounties to encourage men to join voluntarily.

In 1779, Massachusetts recruits for the Continental Army received a bounty of thirty pounds and one hundred acres of land in exchange for nine months of service. Massachusetts assumed the levity of the crisis and the two month enlistments were enough to entice volunteers for the expedition. Unable to raise enough volunteers, commanders resorted to impressment.

The York County commander reported that several of his soldiers had to be brought by force of arms. The Adjutant General, Jeremiah Hill, reported similar issues with men *"sulking off"* to avoid service.

The demands of the expedition, to be met in less than two weeks, simply overwhelmed the draft system used to dealing with filling out smaller quotas. Pressed to fill their quotas, the counties did their best. The rapid demand for troops broke the traditional method of volunteerism.

Adjutant General Hill summed up the minimum qualifications for the soldiers of the expedition, "if they belonged to the train band or alarm list they were soldiers, whether they could carry a gun, walk a mile without crutches or only compos mentis sufficient to keep themselves out of fire and water."

The patriotic zeal that brought men to Lexington in 1775 disappeared to the realities of protracted war.

Brig.-Gen. Lovell was ordered, June 26, 1779, 2 to hold himself in readiness to take the command of twelve hundred militia, with one hundred of the artillery, to march, at the shortest notice, to Penobscot. Orders were issued to the Board of War to fit out a fleet immediately, by obtaining the loan of the frigate *T Warren* " and sloop " *Providence*," *Continental* vessels, the former a fine new ship of 32 guns, and the latter a sloop of 12; to buy, hire, or impress private armed ships as many as might be necessary; also, to provide transports. Extraordinary inducements were also offered for seamen.

Orders were also issued to the ordnance, commissary and quartermaster s departments, to furnish what supplies might be needed from their stores. Ammunition, provisions, and supplies of all kinds in abundance, were ordered, that the expedition might not be lacking in these particulars. Yet, notwithstanding the utmost endeavors of the authorities, difficulties were encountered from the outset. The supplies so liberally ordered were not forthcoming; and Gen. Lovell writes to the Council, under date of July 2, stating that:

so great is the difficulty in obtaining the necessaries under the ordinary method, that the operations of the expedition must be retarded thereby, and requests additional powers, which are granted.

July 7, Peleg Wadsworth, Adjutant-General of the State, an officer of high repute, who had seen much service, was chosen unanimously by the Council to the second position under Gen. Lovell, to serve as engineer, with rank of Brigadier 2 . On the 8th, Lieut-Col. Paul Revere was appointed to command the train of artillery, and, on the 9th, Dr. Eliphalet Downer, Surgeon-General of the expedition; and, so great was the despatch, that the General received his orders to embark his artillery on the 12th, and on the 15th the expedition was in Nantasket Eoads, ready to sail; but, from various causes, it did not put to sea until the 19th. The letter of instructions to the commanders was very full and explicit.

The fleet had been placed under the command of Dudley Saltonstall, Esq., of]^ew London, an officer of some repute, and then in command of the Continental frigate " Warren," which position was thought, perhaps, to entitle him to the command of the squadron.

This consisted of the ships "Warren," 32 guns; the "Hampden" (the New Hampshire contingent, which joined them at Townsend), and "Hector," of 22 guns each; the "Gen. Putnam," "Vengeance," "Monmouth," "Black Prince " (of Salem, which joined the fleet off Portsmouth), "Hunter" and "Charming Sally," each 20 guns; "Sky Rocket," brigs "Hazard," "Pallas" and " Defence," of 16 guns each; "Active" and " Tyrannicide," of 14 each; the "Diligence" and sloop "Providence," of 12 each; sloop "Charming Polly" and schooner "Hannah," of each; mounting in all 324 guns, and manned by more than of two thousand men, with upwards of twenty transports probably the strongest and finest naval force furnished by New England during the Revolution; and the total cost of the expedition, as seen in the general account, was 1,739,174

The expedition was the work of Massachusetts, notice only being given to the Continental authorities, who consented thereto, furnishing aid and counsel, and, too late to be of any avail, a strong reinforcement. The burden imposed upon the colony was a heavy one, and its disastrous result aggravated the burden. The fleet was ordered to rendezvous at Townsend, where the land forces of York and Cumberland, six hundred men each, were to meet them, and where everything was supposed to be in readiness for an immediate departure.

The fleet reached Townsend on the 21st, where the General and his w family " were hospitably entertained by Rev. Mr. Murray, whose place " was a much Genteeler seat than was by most persons expected to be found in this part of the country. "Very agreeably & sociably treated by the worthy clergyman." Mr. Murray was afterwards persuaded by the General to accompany the expedition, and was sent as bearer of despatches to Boston. The General speaks of him in very high terms.

Upon examining the returns of the troops, they were found to be deficient over one-third, or five hundred of the fifteen hundred ordered, which included three hundred from Lincoln. They were, notwithstanding, ordered up for review, and to embark immediately; of the men sent forward, a large part were wholly unfit for service. General Wadsworth says, " at least one- fourth part appeared to me to be small boys and old men, unfit for service."

Adjutant-General Hill says, *"The difficulty in collecting troops was so great that I recommended martial power, as they were legally detached soldiers, and subject to martial law, which was done, and the quota partly filled in that way. Collected four hundred and thirty-three, rank and file, and embarked them for Townsend. Some were old men, some boys, and some invalids; if they belonged to the Train Band or Alarm List, they were soldiers, whether they could carry a gun, walk a mile without crutches, or only composmentis sufficient to keep themselves out of fire and water."* *The Cumberland recruits were of much the same general character. These facts were represented to the General, who wrote at once to the several brigadiers to fill up their quotas immediately. Some of these subsequent levies reached the Penobscot just previous to the final catastrophe."*

The adjutant says further, in relation to their equipments: *"most of them had arms, but many were out of repair, little or no ammunition, and most of the officers and men quite unacquainted with any military manoeuvre, and even the manual exercise."*

Major Todd states 1 that he *" received orders on the 2d July to repair to York County, to receive the troops raised there. Arrived at Wells on the 6th, and after the most urgent endeavors, consulting with General Frost and the colonels of the respective regiments,"* he had received, up to the 13th, not more than sixty men, some of whom were brought in by force of arms; on the 14th he had ready to march one hundred and thirty men, and the next day set out for Casco Bay.

These testimonies are necessary in order to under stand the material with which the General had to work, and the difficulties that beset his way from the very outset. Even recruits of this class were deficient in number fully one-third.

While at Townsend, General Lovell, pursuant to instructions, held an interview with chiefs of the "Bridgewalk" Indians, and finding them apparently friendly and ready with their promises, he supplied them with necessaries. It appears from their subsequent conduct that the friendship was real, and the promises fully carried out, the expedition receiving valuable aid, not only in men for the army, but also in service as guides; a large number being engaged in the military operations.

PEXOBSCOT EXPEDITIOX.
OPENING THE CAMPAIGN.

THE arrangements were fully completed on the 23d, and on the morning of Saturday, the 24th, the armament set sail from Townsend, with a fair wind, arriving the same evening at the mouth of Penobscot Bay, and cast anchor under the Fox Islands. They had observed numerous fires along the coast from point to point as they proceeded, a fact very unusual at this season of the year. These were, nodoubt, the work of British emissaries to give warning of the approach of the enemy. Here they were joined by some Penobscot Indians, who proceeded with them. They had been tampered with by General McLean, but had refused his offers.

On Sunday morning, Captain Mitchell of Belfast having been engaged as guide, the fleet proceeded up the bay, the transports coming to anchor under the bluff at " Bragaduce," about seven o clock in the evening, under cover of ship "Charming Sally" and brigs "Hazard" and "Active"; several of the ships saluting the small battery at the water-side with broadsides as they passed. A landing was at once attempted under the bluff, which was thickly covered with brush and trees, but the sea being so rough, on account of the high wind, that there would be danger that the first division might be cut off before the second could be brought to its support, counter-orders were issued, which reached the first division just as they received the fire from the enemy, who lay concealed among the brush, where they could not be seen. They re-embarked with the loss of one Indian killed.

News of the intended expedition had reached General McLean on the 18th, to which he paid little attention.

On the day following, the intelligence was partially confirmed, and the work on the fortifications renewed with the greatest vigor; the men working night and day, assisted by about one hundred of the inhabitants, who served as volunteers, clearing off the wood, for which they received the thanks of the General. Notwithstanding, however, their most strenuous endeavors, when the enemy arrived off the harbor they were in no posture for defence, and were greatly disheartened.

On the appearance of the American fleet "*the seamen at work on the fort were recalled; the Albany, North and Nautilus formed in close order across the entrance of the harbor, just inside of the rocks on Magabagwaduce Point, and the point off Bank s Island, afterwards called Nautilus, or Cross Island, giving berth for three transports out of line of fire. The troops were encamped about half a mile from the works. The well bastion was not yet begun, nor that of the seamen quite finished. Now the works were put into the best defensible condition, some guns mounted, the army in garrison, and gunboats watching the enemy,*" while urgent despatches were sent to Halifax for immediate reinforcement.

On the 26th, the first division was ordered by General Lovell to make a feint of landing on the bluff head of Maga-Bagaduce, and the marines to attack the enemy upon Bank s Island, a position commanding the shipping in the harbor and also one of their batteries. The attempt was entirely successful. The marines made good their landing, secured the position, driving the enemy from the island, capturing at the same time four cannon and some ammunition, without the loss of a man. The position was immediately taken possession of by General Wadsworth with the first division, which had left its feint for the purpose, although in making the landing, a chain-shot from the enemy s shipping sunk one of the boats, and "*the worthy*

Major Littlefield," with two men, was drowned.

Entrenching tools were ordered on shore at once, an embankment thrown up, and a battery mounted, consisting of two eighteen- pounders and one twelve, in addition to a brass howitzer and a field-piece. The retreat of the British was so precipitate that they left their tents standing, and their flag as a trophy fell into the hands of the marines, who presented it to General Lovell. The post was left in charge of Captain Barker, with Captains Johnson and Edmunds and a detachment of troops. This movement compelled the British to withdraw their ships to a position farther up the harbor.

During this time the fleet under Commodore Saltonstall had kept up, at intervals, a desultory cannonade upon the enemy with very little result. There appeared to be a disposition on the part of the Commodore to avoid any offensive movement, and to keep his fleet as far from danger as possible. The effect of this action, or rather want of action, was such as to cause the greatest dissatisfaction and disgust among the officers of the fleet, who did not allow this feeling to conceal itself or die for want of expression.

As early as the 27th, a circular, certainly not at all ambiguous, signed by upwards of thirty of the lieutenants and masters, expressing the feelings of nearly all of the officers under his command, did not, as the event proved, produce the desired effect, for, while assenting ostensibly to the necessity of an immediate attack by the fleet upon the enemy s insignificant force, obstacles and delays were always the order of the day, and the work was not attempted.

On the afternoon of this day, when the Commodore must have felt the full force of this composing draught, a council of naval and land officers was held on board the " Warren," and the determination reached, to land upon the peninsula now in possession of the enemy, to obtain a permanent foothold, and to dislodge them if possible.

Accordingly, before light on Wednesday morning, all the troops were ordered into their boats, and a little before sunrise were formed, and, with hearty cheers, pushed for the shore, under cover of the guns from the fleet, intending to land under the high, precipitous bluff forming the south-western base of the peninsula, here nearly two hundred feet in altitude, and of nearly perpendicular ascent. The almost inaccessible nature of the shore, had led the British to believe that no attempt to land at this place would be made; therefore, no protective works had been erected; the steep bank and the thick brush affording sufficient protection for the troops necessary for its defence.

Under this cover, some three hundred of the enemy were posted, who, as soon as the boats struck the beach, opened their fire. The American force was formed in two divisions, the marines, about one hundred and fifty of whom were in the engagement, with a part of Colonel McCobb's militia on the right, the remainder in the centre, General Lovell landing with the latter.

Notwithstanding the extreme difficulty of the ascent, and the enemy s fire directly in their faces, the troops pushed on with the greatest intrepidity, although with but little order, scaled the heights, swept the foe before them, and captured a position upon the bluff which was of the highest importance, since it gave them a point from which future operations against the fort could be conducted with the greatest advantage.

The hard fighting was upon the right, the marines suffering severely, while the other division, closing in with too much precipitation, drove the enemy from the ground, and enabled them to escape. The fight lasted but twenty minutes, and considering that the attacking force was composed of undisciplined militia, most of whom were never before in action; the ascent almost too difficult to be undertaken unopposed, made in the face of a strong party of veteran troops, it may be fairly set down as one of the most brilliant exploits of the war.

Says General Lovell, *"When I returned to the Shore, it struck me with admiration to see what a Precipice we had ascended, not being able to take so scrutinous a view of it in time of Battle; it is at least where we landed three hundred feet high, and almost perpendicular, & the men were obliged to pull themselves by the twigs & trees. I don't think such a landing has been made since Wolfe."*

The loss of the Americans was fourteen killed and twenty wounded, including the *"brave Major W T Welch of the marines, and Capt. Hinckley of the Lincoln militia,"* while that of the enemy was fifteen killed, three wounded and eight prisoners. The American loss was greatly exaggerated by the enemy, some accounts making it as high as one hundred; but the report of Loveirs Journal

General Lovell, from which these figures are taken, is undoubtedly correct.
Orders were immediately given to secure the position, which was within point-blank range of the enemy's fort, by intrenchments and a battery. This compelled them to abandon their battery on South-east Point, leaving behind them three six-pounders, and also to with draw their shipping to a point inside, commanded t by their batteries, and " out of reach of our shot."

PROGRESS OF THE SIEGE.

The 29th, a new battery was erected by the Americans, about sixty rods in advance of their former lines, and but a quarter of a mile from Fort George, mounting two eighteen-pounders, one twelve-pounder and one howitzer, which were ready for duty on Friday and opened fire, the fleet threatening at the same time hostile measures. At this demonstration, the enemy sunk most of their transports and retired with their artillery to the fort, which was the only ground now held by them, except a small redoubt that protected their shipping.

While this movement was in progress, fatigue parties were engaged in strengthening the works on the heights, also in making a covered way across the isthmus connecting with the main, and in clearing a road in case a retreat should become necessary. Cannonading was carried on for several days, between the fort and the ships, assisted by the batteries, but generally with out result. A packet from Halifax, taken by the fleet, was brought in, but the despatches had been destroyed; the only news, which was obtained from prisoners, being the constant expectation of a reinforcement by the enemy, and the fact that they were at work day and night, strengthening their positions, every day's delay being of great value to them.

About two o clock on Sunday morning, August 1, General Lovell detached General Wadsworth with about three hundred men, a part of whom were sailors and marines, to capture the redoubt that covered the enemy s shipping and commanded the harbor.

They marched forward in good order until they received the fire from the garrison, when they broke ; "*a few, how ever, nothing daunted, pushed bravely forward and forced the battery, but were obliged to destroy it, as it was commanded by the enemy s main fort. They killed five of the enemy and captured eighteen, destroying their stores, with a loss of four missing and twelve wounded; among the latter was Major Sawyer.*"

A proclamation which had been issued by General Lovell to the neighboring inhabitants on the 29th, to counteract the British influence in that direction, was quite successful, and the people were beginning to come in freely in consequence. The greatest harmony and zeal existed among the troops, who were very active in carrying forward the plans of the General, not withstanding they were suffering severely from a storm that came up, having shelter.   Rev. Mr. Murray, "*who has distinguished himself as a citizen and a soldier, who has undergone the fatigues of my camp and finding it necessary to despatch a courier has voluntarily offered his services,*" was sent to Boston for reinforcements and supplies.

Commodore Saltonsatall Refuses to Attack

During all this time General Lovell had been using his utmost endeavors to persuade the Commodore to go in with the fleet and destroy the few ships of the enemy remaining in the harbor, when the fort could be attacked with good prospect of success; but the Commodore declined, unless the General would storm the fort as the latter did not feel himself strong enough to do without the aid of the marines to co-operate.

At one time the Commodore answered the request by pointing to the three-gun battery, destroyed on Sunday morning; at another he urged that his ships might suffer, and as there was no place at which to refit, he might fall a sacrifice should a reinforcement arrive in aid of the enemy.

And thus there was delay upon delay, and every day was a golden opportunity improved by the enemy. Nor was General Lovell the only one to complain of the inactivity and Avant of enterprise of the Commodore; his own officers were equally dissatisfied, as is fully shown by the letter addressed to him on the 27th, already mentioned.

They were almost unanimously in favor of attacking the ships at once. Colonel Brewer, who was in the fort only the day before the arrival of the fleet, told the Commodore that *"he (the Commodore) could silence the vessels and the battery in half an hour, and have everything his own way."* l was answered by an oath, and *"I am not going to risk my shipping in that d d hole."* To Captain Titus Salter, of the "Hampden," who ventured a similar suggestion, he replied by threatening to reduce his ship to a bread transport.

Owing to the refusal of Commodore Saltonstall to go in with his ships, it became necessary to adopt other measures to act against the enemy s vessels, and General Wadsworth was sent on the 3d with a detachment to erect a battery upon the main, opposite their anchorage, to drive them away. They landed at Swet's Cove, and with the aid of some of the seamen from the " Hazard " and w "Tyrannicide," placed in position a battery, mounting one eighteen-pounder, one nine-pounder, and one field-piece, and opened fire, but with little effect, the distance was so great, being a mile and a quarter; and the General writes sadly in his journal of the 4th, *" it is all the Army can do they have tried their best."*

For several days the progress of the siege was a succession of cannonades, alarms and fatigue duty, principally the latter, the outworks of the Americans being within musket-shot of the fort, and the whole army in the woods, within point-blank range.

On the 6th, General Lovell again wrote Commodore Saltonstall, desiring to know *"if he would go in and destroy the three sloops of War of the enemy,"* with the same result; and again the General writes the Council by Major Braddish, urging immediate reinforcements.

There was little change in the position of affairs for several days. Occasional skirmishing, cannonading from the batteries and shipping upon the enemy's works, but with no important results. In the mean time, it was becoming more and more apparent that a crisis was approaching when some decisive movement must be made, or the expedition abandoned.

The continued inaction, with stormy weather, which was causing great loss of ammunition and provisions, and want of proper shelter was having its effect upon the men, who were fast becoming demoralized.

To test the temper and discipline of the troops, Gen. Lovell, on the 10th, ordered out a strong skirmishing force, under Adjt. Gen. Hill, consisting of six hundred men, volunteers, if possible. The impression having gone abroad that a general assault was intended, it was with extreme difficulty that four hundred were obtained; Col. Mitchell filling his quota of two hundred, after great exertion, by includindg old men, boys and invalids.

Col. McCobb succeeded in raising about one hundred and fifty, and Maj. Cousins, seventy-five, but twenty of them deserting during the night, and thirty more having been detached to look up the fugitives, he could furnish none; the remainder,

about fifty, were made up from the new levies.
" Col. Mitchell's officers were so terrified that they complained of his nomination, and even drew lots as to who should go."

On the following day, they were thoroughly tested in the field with the enemy; but with such results that the General did not dare to undertake any important movement. At the same time, the Navy Board, informed of the slow progress of the siege, and the want of co-operation on the part of the fleet, wrote to the Commodore, Aug. 12, complaining, in very strong terms, of his inaction and backwardness in not attacking and destroying the British shipping, when, by general acknowledgment, it was in his power to do so, and directing him to do it at once.

On the llth, the General had written to the Commodore a very severe letter which was found by the enemy on a captured transport, and afterwards published, in the following terms:

SIR: In this alarming posture of affairs, I am once more obliged to request the most speedy service in your department; and that a moment be no longer delayed to put in execution what I have been given to under and was the determination of your last council.

The destruction of the Enemy s ships must be effected at any rate, although it might cost us half our own; but I cannot possibly conceive that danger, or that the attempt will miscarry. I mean not to determine on your mode of attack, but it appears to me so very practicable, that any farther delay must be infamous; and I have it this moment, by a deserter from one of their ships, that the moment you enter the harbor they will destroy them, which will effectually answer our purpose."

The idea of more batteries was reprobated, having been sufficiently tried; besides, "that would take up dangerous time." He expresses his ardent desire to co-operate with the fleet in active operations; that the army had reached the limit of its power; the probability of a speedy reinforcement of the enemy necessitating instant action or the disgrace of losing their ships, the retreat of the army being secured. He continues : " I feel for the honor of America, in an expedition which a nobler exertion had, long before this, crowned with success; and I have now only to repeat the absolute necessity of undertaking the destruction

In the meantime, councils had been held, nearly every day, of land or sea forces, or both combined; but the fact that the Commodore was averse to action, declining to risk an attack for fear of damaging his vessels, and that a large part of his captains, their ships being private property, shared the same feelings, and that there was but little show of prize-money, produced conflicting opinions, and prevented decisive results.

On the 7th, an incident of a ludicrous character occurred, which exposed the actors to no small amount of disgrace. The Commodore, with five of his captains, while reconnoitering in an open, unarmed boat near the enemy, was discovered by them, who, guessing their character, immediately fitted out a strong party, in eight boats, and gave chase.

The pursuit was so sharp that the Commodore and his company, in order to escape capture, ran their boat on shore, and took to the bush. The boat fell a prize to the enemy, while the officers, after remaining on shore all night, succeeded in reaching the fleet the next morning.

THE DEFEAT.

ON the 12th, however, Gen. Lovell came to the determination to take up such a position as should compel the Commodore to move, although his force was, at this time, really inferior to that of the enemy, being but about nine hundred men, including the train of artillery and volunteers, although expecting reinforcements from Cols. Allen and Foster, and hoping for some, also, from the Continental government.

Accordingly, on the afternoon of the 13th, he proceeded, with four hundred men, to the rear of the enemy s position, and took -post where he could operate to the best advantage. The move was a dangerous one, but gave fair promise of success. Gen. Lovell immediately despatched intelligence of his action to Commodore Saltonstall, who had always insisted that the army should attack the fort before the fleet should enter the harbor. In the first return made during the expedition, July 20, eight hundred and seventy-three men reported fit for duty; the second, July 31, eight hundred and forty-seven men ; the third, Aug. 4, seven hundred and sixty- two ; the fourth, Aug. 7, seven hundred and fifteen ; and the fifth, and last, nine hundred and twenty-three, with one hundred and thirty-eight on the sick-list, one hundred and forty-four on command, and eight on furlough ; two companies having joined Col. McCobb since the previous return, the artillery and volunteers being included only in the last return. [State Archives, vol. 145, pages 48, 60, 66, 83 and 101.]

Arrangements had already been made to send Col. Jackson, with four hundred regulars, also a naval contingent, with additional supplies, which were, however, too late to be of any avail, the expedition having been defeated about the time the reinforcements sailed from Boston.

They at once weighed anchor, but had no sooner made sail than report was brought that a fleet was entering the bay. Word was instantly conveyed to the General, who, without loss of time, returned to his previous position. At twelve in -the night, intelligence came from the Commodore that the strange fleet were ships of force, and British.

Retreat

Orders were given for an immediate retreat, which was effected in good order, and without loss. The batteries were dismantled, and the artillery re-embarked on the transports, the fatigue parties, with the entrenching tools, and every other article of value, were on board by daylight, the troops by sunrise, and orders were given to proceed up the river, under command of Gen. Wadsworth.

The only articles not brought off were two eighteen-pounders and one twelve-pounder, on an island at the entrance of the harbor, under the care of the officers of the navy. The General used every effort to secure these, but the time was too short, and the covering ships had withdrawn.

The transports immediately proceeded up the bay; but when they reached the mouth of the river, about two leagues away, the breeze falling and the tide being on the ebb, they cast anchor. The General, meanwhile, took a barge, and waited upon the Commodore, to try and induce him to offer what resistance was possible to the British, and thus enable the transports, with the troops and stores, to escape to some point on the river above, where a stand could be made, and the fleet, perhaps, saved; but, on reaching the "Warren," they were told that it was determined to run up the river, and the ships were even then getting under way for the purpose, a light breeze beginning to be felt, while the enemy s first division was coming in under full sail, distant about two miles.

Learning the determination of the Commodore, Gen. Lovell, after expressing his surprise at the movement, embarked on board his boat to return to his troops; but the breeze increased so rapidly that the ships-of-war soon came up with them, and the General was taken on board the "Hazard."

Consternation and Confusion

From this time forward, there was but one continued scene of consternation and confusion. Finding that they were to receive no support from the armed vessels which were only doing their best to make good their own escape, the transports immediately proceeded to get under way, just feeling the breeze; but, being now astern, close to the enemy, and finding that they must inevitably fall into their hands, nothing was thought of by the crews but as speedy escape as possible to the shore, and hardly an attempt was made to save anything. Some were run on shore, some anchored, some abandoned with all sails set, and most set on fire.

Officers were despatched by Gen. Lovell to the shore to collect and take charge of the troops; but so great was the panic, so convenient the woods and the approaching night, that but few could be found; the greater part, thinking that nothing further was expected of them, made the best of their way, singly or in squads, towards the Kennebec, where the most of them arrived, after nearly a week s fatigue, suffering greatly from exposure and hunger, some of them tasting no food for several days.

The ships-of-war were in no better condition than the transports, simply flying into a trap whence they could be taken at leisure. The General, fearing their destruction, hastened to secure their safety, urging that a line be formed across the river, and a defence made at some point where that could easily be done, offering to support them with the troops that remained; but upon application to the Commodore to know if any measures had been concerted for their security, he found him wholly undetermined and irresolute completely unmanned.

The hostile fleet consisted of seven sail, one two decker, two frigates, two sloops-of- war, with two smaller vessels, carrying two hundred and four guns and fifteen hundred and thirty men, under Sir George Collier. This force, with the three sloops-of-war already in the harbor, presented too strong a force to be successfully contested in the open sea; but in a river, offering so many points of easy defence as the Penobscot, the result was shameful, since the American force was yet numerically much superior to that of the enemy in guns and men, although inferior in weight of metal and tonnage.

Sir George made no delay, but proceeded at once to attack his foe, and his boldness had the desired effect, producing, as has been seen, the panic that resulted in the total destruction of the fleet. The *Hunter, Hampden and Defence*, in attempting to reach the sea by the western passage, round the head of Long Island, were intercepted, the two former captured, and the latter run into an inlet and set on fire. The remainder of the fleet fled before the enemy up the river and were all set on fire and blown up at various points.

Paul Revere Abandons his Ship

The ordnance brig, on board of which was all the artillery and ammunition, with the troops of Lt. Col. Revere (he having gone on shore at Fort Pownal), the sole dependence of the army in case a stand should be made, was deserted, but cleared herself from the transports/and made her way alone up the river for several miles, but was then boarded, set on fire, and burned with all her contents.

The destruction was complete, but two or three of the vessels falling into the hands of the enemy. In relation to the last act in this disgraceful drama, the General writes: *"The Transports then again weigh d anchor, and to our great mortification were soon followed by our fleet of men of war, pursued by only four of the enemy s ships, the ships of war passed the transports, many of which got aground, and the British ships coming up the soldiers were obliged to take to the shore and set fire to their vessels. To attempt to give a description of this terrible day is out of my power. It would be a fit subject for some masterly hand to describe it in its true colors ; to see four ships pursuing seventeen sail of armed vessels, nine of which were stout ships transports on fire men of war blowing up provisions of all kinds, and every kind of stores on shore (at least in small quantities) throwing about, and as much confusion as can possibly be conceived."*

After the destruction of the fleet and the dispersion of the troops, excepting a few remnants which he placed in charge of Gen Wadsworth, Gen Lovell proceeded up the river to treat with the Indians, among whom there appeared great uneasiness, some outrages having been already committed by them, which had excited the apprehensions of the inhabitants.

He succeeded in Major Todd's Report, " Chronicle and Advertiser," Boston, Sept. 30, 1779 : " *In the pursuit of the rebel fleet up the Penobscot, the King s fleet were obliged to come to anchor, on account of the rebels having moored a sloop in the channel and set her on fire ; otherwise their whole fleet would have been captured She was soon towed out of the way, when the rebels blew up and set fire to most of their shipping."*

This accomplished, he took his departure for the Kennebec region under the guidance of some of the friendly Indians, where he arrived safely, but after suffering much fatigue. He here received orders to take post at some point in the eastern part, the best suited, in his judgment, for its protection, with Col. Jackson and the force under his command, and was also empowered to call upon the militia of that section for such reinforcements as he should find indispensably necessary.

Having completed his arrangements and settled the military affairs of that part of the Province as well as circumstances would admit, he proceeded to Boston, where he arrived about the 20th of September.

The entire failure of this important expedition, of which so much had been expected, and upon which had been expended such an amount of money from the already depleted treasury of the Province, caused immense excitement; and the pressure was so great that the General Court felt called upon to investigate the matter.

Investigatory Committee Appointed
Accordingly, on the 9th of September, the General Assembly appointed a Committee to look into the causes that produced it, to give a most careful examination and report the result. This Committee consisted of Generals Michael Farley , and Jonathan Titcomb, Col. Moses Little, Major Samuel Osgood and James Prescott, Esq., to whom were joined from the Council Generals Artemas Ward and Timothy Danielson, Hon. William Sever and Francis Dana, Esq.

This Committee organized, with General Artemas *"Ward, chairman, and after a most thorough hearing, having examined more than thirty witnesses from the naval and military departments of the expedition, on the 7th of October made the following report, in the form of interrogatories and answers"* .

" 1st Question. Is it the opinion of this Committee that they have made sufficient inquiry into the causes of the failure of the late expedition to Penobscot? "

Answered unanimously, Yes.

" 2d Question. What appears to be the principal reason of the failure? "

Answer, unanimously, Want of proper spirit and energy on the part of the Commodore.

"3d Question. Was General Lovell culpable in not storming the enemy s principal Fort according to the requirement of the Commodore and Naval Council, who insisted upon that as the condition of our ships attacking the enemy s ships, when at the same time the Commodore informed him in case of such an attack he must call the marines on board their ships (the last was not made a part of the condition by the Naval Council) ? "

Answer, unanimously, No.

"4th Question. What, in the opinion of this Committee, was the occasion of the total destruction of our fleet? "

Answer. Principally the Commodore's not exerting himself at all at the time of the retreat in opposing the enemy s foremost ships in pursuit.

"5th Question. Does it appear that Gen. Lovell throughout the expedition and the retreat acted with proper courage and spirit?"

Answer, unanimously, Yes, it is the opinion of the Committee had he been furnished with all the men ordered for the service, or been properly supported by the Commodore, he would probably have reduced the enemy.

"6th Question. Does it appear that the Commodore discouraged any enterprises or offensive measures on the part of the fleet?"

Answer, unanimously, Yes, and although he always had a majority of his Naval Council against offensive operations, which majority was mostly made lip of the commanders of private armed vessels, yet he repeatedly said it was matter of favor that he called any Councils, and when he had taken their advice he should follow his own opinion.

As the naval commanders in the service of the State are particularly amenable to the Government the Committee think it their duty to say that each and every of them behaved like brave, experienced, good officers throughout the whole of the expedition.

" 7th Question. What was the conduct of Brigadier Wadsworth during his command?"

Answer. Brigadier "Wadsworth (the second in command) throughout the whole expedition, during the retreat and after, till ordered to return to Boston, conducted with great activity, courage, coolness and prudence.

The Committee find that the number of men ordered to be detached for this service were deficient nearly one-third. Whether the shameful neglect is chargeable upon the Brigadiers, Colonels, or other officers whose particular duty it might have been to have faithfully executed the orders of the General Assembly, they cannot ascertain. Oct. 7, 1779.

It appears from the records of the expedition that a warrant for a court-martial for the trial of Dudley Saltonstall was issued at the same time the Committee of Inquiry was ordered, September 7, to meet on board the *Deane* frigate on the 14th.

Hon. Mr. Sever not present at all at the enquiry, and the Cols, Prescott and Little absent when this report was made.

Court Adjourns for Two Weeks
On the 14th the court met and adjourned to the 28th of the same month, at the request of the Naval Board, and in accordance with a resolve of the General Court, to see what action the latter body would take in relation to the matter, the Commodore also complaining that hasty action would greatly prejudice his cause.

The most careful search among all known sources of information fails to discover any further traces of this court- martial, although several of the accredited histories, and general tradition, state that he was cashiered and pronounced forever incapacitated for holding governmental office. (The records of this proceeding may have been filed at Washington in the Navy Department and destroyed when the public buildings were burned by the British in the War of 1812-15.)

That such was the result in his case there can be little doubt, from the finding of the Court of Inquiry, and from the fact that he disappears from that time, and is never heard of afterwards in the public records, while the other officers prominent in the expedition retained their positions, and the confidence of the authorities, including Lt. Col. Revere, who was censured for his conduct while in that service.

Brigadier General Solomon Lovell Lauded

The Penobscot expedition, while it reflects lasting disgrace upon the one chief delinquent (whether acting from cowardice, or bribery, or both, it has been impossible to determine), casts no discredit upon the commander of the land forces, but leaves him with an untarnished reputation, as a brave, patriotic and skilful general.

It has been seen that, from the outset, difficulties crowded in his way. Delay followed delay in the fitting out and sailing of the fleet from Boston. The short complement and inferior quality of the men provided by the officers ordered to furnish them, on their arrival in Maine, and the want of co-operation on the part of the Commodore, were obstacles not easy to be overcome, even by acknowledged genius.

It may be objected that he did not act with sufficient promptness and energy, which, had he done, the fort would have fallen upon the first attack; but the disheartened condition of the garrison and the weakness of the works could hardly have been known to the General; on the contrary, he knew the almost total want of discipline of his own force, which report had exaggerated to the enemy to four times its real number; and he knew, also, that the force opposed to him was nearly equal to his own in numbers, besides being veteran troops ; that they had been in active preparation for more than a month; and he had strong reason to believe that they were fully ready for his attack, while his main reliance was the fleet. Had the enemy s ships, whose fire covered their fort, been destroyed, his work would have been plain and comparatively easy.

The landing effected on the 28th July was one of the most brilliant exploits of the war, reflecting the highest credit upon him who planned and executed it; nor, had it not been for the subsequent misfortune, would it have suffered in comparison with the more widely celebrated capture of Stony Point by Wayne, a nearly contemporaneous action.

It has been shown how persistent were his endeavors in urging the Commodore up to his duty, and his efforts in taking advantage of every circumstance to reduce his enemy on the land within the smallest possible compass; also, how well he succeeded in the latter by confining him at last to the walls of his fort; and the report of the Committee of Inquiry of the General Court will be fully sustained by any one who looks carefully through the voluminous reports of the expedition on the files of the State department in Boston.

Court's Finding

After hearing the whole report, from which the above are but quotations, the General Court adjudged –

"that Commodore Saltonstall be incompetent ever after, to hold a commission in the service of the state and that Generals Lovell and Wadsworth be honorably acquitted."

Two More Attempts to *Establish New Ireland* after the Revolutionary War is Won

Another Attempt by the British to Establish *New Ireland*

War of 1812 and the Battle of Hampden

This battle was an action in the British campaign to conquer that part of Maine north of the Penobscot River Maine and remake it into the colony of *New Ireland*. During the War of 1812 Sir John Sherbrooke led a British force from Halifax, Nova Scotia to establish New Ireland, which lasted until the end of the war, eight months later. The subsequent retirement of the British expeditionary force from its base in Castine, Maine to Nova Scotia ensured that eastern Maine would remain a part of the United States. Lingering local feelings of vulnerability, however, would help fuel the post-war movement for statehood for Maine. The withdrawal of the British represented the end of two centuries of violent contest over Maine by rival nations (initially the French and British, and then the British and Americans).

Capture of Castine

On August 26, 1814, a British squadron from the Royal Navy base at Halifax moved to capture the Down East coastal town of Machias. The force consisted of five warships: HMS *Dragon* (74), HMS *Endymion* (50), HMS *Bacchante* (38), HMS *Sylph* (18), a large tender, and ten transports carrying some 3,000 British regulars (elements of the 29th, 60th, 62nd, and 98th regiments and a company of Royal Artillery).

The expedition was under the overall command of Sir John Sherbrooke, who was then the lieutenant governor of Nova Scotia. Major General Gerard Gosselin commanded the army and Rear Admiral Edward Griffith Colpoys controlled the naval elements.

Re-establish New Ireland

The intention of the expedition was clearly to re-establish British title to Maine north of the Penobscot River, an area the British had renamed *"New Ireland"*, and open the line of communications between Halifax and Quebec. Carving off *"New Ireland"* from New England had been a goal of the British government and the colonies of New Brunswick and Nova Scotia *("New Scotland")* since British Brigadier General Francis McLean conquered Maine in 1779 during the American Revolution.

En route, the squadron fell in with HMS *Rifleman* (18), and learned that the USS *Adams* (28), commanded by Captain Charles Morris, was undergoing repairs at Hampden, on the Penobscot River. Sherbrooke changed his plan and headed for Castine at the mouth of the Penobscot. He rendezvoused off Matinicus Island and added HMS *Bulwark* (74), HMS *Tenedos* (38), HMS *Peruvian* (18), and the schooner (18) and HM Schooner HMS *Pictou* (14) to his force. The complete force entered the cove at Castine on September 1.

The local American militia melted away at the sight and a 28-man contingent from the U.S. Army under Lieutenant Andrew Lewis destroyed their four 24-pounders, blew up their magazine and withdrew to the north trailing a pair of field pieces.

As the first order of business, Sherbrooke and Griffith issued a proclamation assuring the populace if they remained quiet, pursued their usual affairs and surrendered all weaponry; they would be protected as British subjects. Moreover, the British would pay fair prices for all goods and services provided.

Belfast Occupied Next

Major General Gosselin then crossed the bay with most of the 29th to occupy Belfast and protect the left flank of the major operation to follow. Locals did not challenge the occupation, although some 1,200 militiamen gathered three miles outside Belfast to await developments.

Expedition up the Penobscot River

Rear Admiral Griffith assigned Captain Robert Barrie the task of going after the American's war ship *Adams* under repair up the Penobscot River in the town of Hampton. Barrie proceeded up the Penobscot with the *Dragon, Sylph, Peruvian,* the transport *Harmony* and a prize-tender. The ships carried an armed contingent of some 750 men drawn from the four participating regiments, the artillery company, and some Royal Marines. During the war, Barrie was one of the few British officers in America to acquire a loathsome reputation, which he was about to reinforce.

Battle of Hampden

When Morris entered the river late in August he moved past Buckstown (now Bucksport, Maine) and anchored at the mouth of the Sowadabscook Stream in Hampden on the west bank of the Penobscot some 30 miles inland. Anticipating an attack, he placed nine of the ship's guns in battery on a nearby hill and fourteen on the wharf next to his crippled ship. Morris, commanding a crew of 150, called for help from Brigadier General John Blake, commander of the Eastern Militia at Brewer, Maine.

Blake responded with some 550 militiamen and formed the center of a defensive line running along a ridge facing south, or towards Castine. Lieutenant Lewis showed up with his two dozen or so regulars and two field pieces. Adding a carronade, he went in line to the right or west and commanded the north-south road, the expected route of British attackers.

Late on September 2, Barrie landed his force at Bald Head Cove three miles below Hampden and waited for morning. Early on September 3, in rain and fog, the British moved on Hampden, led by Lt. Colonel Henry John. Skirmishers met with resistance at Pitcher's Brook, primarily from the guns directed by Lewis, but John sent reinforcements and the British stormed across the bridge.

In short order, the full force was in position to continue against the American defensive line on the hill. The sight of the oncoming disciplined Redcoats, bayonets glistening, rattled the untrained militia. The center broke and fled to the woods toward Bangor. Morris on the left and Lewis on the right found themselves in untenable positions.

About to be overrun, Morris spiked his guns and ignited a train leading to the *Adams*. With colors flying, the ship blew up before the British could intervene. Lewis likewise spiked his guns and withdrew to the north. Morris and his navy band made it to Bangor, crossed west through rugged country to the Kennebec River, and around September 9 arrived at their base in Portsmouth, New Hampshire. After two weeks, every sailor reported, not a man missing, a source of great satisfaction for Morris.

At this point, Barrie detailed 200 men to take control of Hampden while he and the balance of his force pursued the Americans in the direction of Bangor. Eighty prominent men of the Hampden area spent a night as prisoners. Most were paroled the next day.

Sacking of Bangor and Hampden

Supported by three of his ships, Barrie entered an intimidated Bangor at midday and called for unconditional submission. Provisions and quarters were demanded and readily turned over *"since the commodore, who was a churlish, brutish monster"*, according to a correspondent, *"threatened to let loose his men and burn the town if the inhabitants did not use greater exertion to feed his men."*

Although Barrie ordered a ban on liquor for his troops, some men managed to acquire brandy by the bucket. Accordingly, Barrie ordered an officer to destroy all liquor in the town. This set off a wave of plundering. Six stores fell to the mob and $6,000 worth of property was damaged. Many citizens fled to the woods. *"We are alive this morning,"* wrote a newspaper correspondent, *"but such scenes I hope not to witness again. The enemy's Soldiery have emptied all the stores and many dwelling houses - they break windows, and crockery, and destroy every-thing they cannot move."*

During the night, the British burned 14 vessels across the river in Brewer. Before the raiders could ignite Bangor vessels, the town's selectmen made a deal. Fearful that the burning would lead to a conflagration, the selectmen offered Barrie a $30,000 bond and agreed to complete four ships on the stocks and deliver them to him in Castine.

Barrie accepted the arrangement and carried away a packet, four schooners and a boat. Before moving back down the river on the 4th, Barrie and John paroled 191 locals considered prisoners, including General Blake. Bangor selectmen estimated that the losses and damages totaled $45,000.

The Bangor diversion did not end the difficulties for Hampden. Barrie decided to spend more time in the town. Redcoats terrorized the village, killing livestock for sport and destroying whatever met their fancy, including gardens, furniture, books and papers. Two vessels moored off the town were burned. The rampage prompted a town committee to appeal to Barrie to treat the place with a little humanity.

His shocking reply summarized his approach. *"Humanity! I have none for you. My business is to burn, sink, and destroy. Your town is taken by storm. By the rules of war we ought to lay your village in ashes, and put its inhabitants to the sword. But I will spare your lives, though I mean to burn your houses."*

Barrie did not follow through on his threat to burn houses, but he did secure a $14,000 bond on several incomplete vessels on the stocks in town. The terms required the completed vessels be delivered to the Royal Navy in Castine by November 1. In the end, the town estimated the value of its losses to total $44,000.

The British then slipped down to Frankfort and demanded considerable livestock and surrender of all arms and ammunition at that place. The locals were slow to comply and before he moved along on the 7th, Barrie promised to return and make the town pay for its delays. The captain did not make good on this threat, and except for some nuisance sniping at the British as they passed Prospect, Maine, the Battle of Hampden was at an end.

Casualties

The British Army loss in the battle was 1 enlisted man killed, 1 officer and 7 enlisted men wounded and 1 enlisted man missing. Four of the casualties were from the 29th Regiment, two from the 62nd Regiment and 4 from the 98th Regiment. The Royal Navy reported 1 sailor from HMS *Dragon* killed. Two British graves in Hampden remain there today, but no details are carved on the stones. These could be the soldiers and sailor killed that day.

American casualties were low, but sources are conflicting. Williamson gives 1 militiaman killed and 11 wounded, with at least two civilians killed by accident. Including the wounded, 84 Americans were taken prisoner. Williamson's data may reflect only the losses to the Hampden militia companies. Captain Barrie could form no estimate, but noted upwards of 30 lying wounded in the woods.

Lt. Col. John states he had no correct number, but reported 30 to 40 killed, wounded or missing. Militia leaders could not confirm how many men actually reported for duty. A list for pay purposes was finally produced but is missing entire companies and states no casualties except for one "Tobias Oakman - killed" (the basis for the "1 killed" that Williamson repeated). Claims by citizens for various compensations were filed for numerous years after the battle without a final tally or surviving documentation.

British evacuation of Castine

Sherbrooke declared "New Ireland" (Eastern Maine) a province of British North America (Canada) and left General Gosselin in Castine to govern it. For the next 8 months (from the fall of 1814 to the spring of 1815), the Penobscot River was essentially an international boundary. That

Hampden and Bangor were on the wrong (American) side might have contributed to their rough treatment. With the signing of the Treaty of Ghent in December 1814, however, the British claim to Maine was effectively surrendered. The British evacuated Castine on April 25, 1815, and the pre-war boundary was restored. The final boundary between the inland, wooded portion of Maine and Canada would remain open to dispute until the Webster-Ashburton Treaty of 1842.

For 8 months (from the fall of 1814 to the spring of 1815), the Penobscot River was an international boundary

---

Aftermath and Consequences

Local memory of this humiliation contributed to subsequent anti-British feeling in Eastern Maine, which would find outlet again in the Aroostook War of 1838-1839. It would also contribute to the post-war movement for Maine's statehood (given that Massachusetts had failed to protect the region) and to the building of a large, expensive granite fort (Ft. Knox) at the mouth of the Penobscot River starting in the 1840s.

General Blake and two other officers (Lt. Col. Andrew Grant of Hampden and Maj. Joshua Chamberlain of Brewer, grandfather of the later Civil War general) were court-martialed in Bangor in 1816 for their part in the defeat. Blake and Chamberlain were both exonerated, but Grant was cashiered.

The elderly Blake was court-martialled first and cleared of charges. He in turn brought charges against his two subordinates in perhaps a move to clear his name. Grant was found guilty of actions unbecoming an officer before the enemy and banned from being re-elected as a militia officer.

One report claims he ran from battle and changed out of his uniform into civilian clothes before eventually being captured and identified.

The Aroostook War of 1842

Maine/Canada border dispute

Sometimes called the Pork and Beans War, was a confrontation in 1838–1839 between the United States and the United Kingdom over the international boundary between the British colony of New Brunswick and the American state of Maine.

The Aroostook War, although devoid of actual military combat, involved heated arguments and negotiations. Neither America nor Britain actually wanted war as it would have greatly interfered with the two nations' trade.

THE WEBSTER-ASHBURTON TREATY, 1842

The disputed land area is marked by hash lines

Daniel Webster negotiated a compromise that was the basis of the *Webster-Ashburton Treaty* in 1842. This treaty settled the Maine-Canada boundary and the boundary between Canada and New Hampshire, Michigan and Minnesota.

Fort Knox, built 1844-1869

Under construction for 25 years to defend against further attacks from British or others

Tensions were high between the British and the Americans given the three attempts by the British to claim what is now "Downeast" Maine even though yet another Peace Treaty, The *Webster-Ashburton Treaty*, had just been negotiated.

The Fort's History
The fort is located on the western bank of the Penobscot River in the town of Prospect, Maine, about 5 miles from the mouth of the river. It was the first fort in Maine built of granite (instead of wood). The purpose of the fort was to defend

against further attacks from the British or others.

Construction funding from Congress was intermittent, and although nearly a million dollars were spent, the fort's design was never fully completed. It is named after Henry Knox, the first US Secretary of War.

The Aroostook War of 1838-1839 revived anti-British feeling and concern over the vulnerability of the region to another attack like that of 1814. Also, the Penobscot valley and Bangor were major sources of shipbuilding lumber.

Interior of the fort looking towards Bucksport

Construction began in 1844 and continued until all masonry fort funding was withdrawn in 1869, with the fort mostly complete except for the emplacements on the "roof" or barbette level. Granite was quarried five miles upriver from Mount Waldo in Frankfort. The fort's overall design was by Joseph G. Totten, the foremost fortification engineer of the Army Corps of Engineers in his day.

Besides the main fort with 64 guns, Fort Knox had two open water batteries facing the river, each equipped with a shot furnace to heat cannonballs sufficiently that they could ignite wooden ships if the ball lodged in the vessel. These furnaces became obsolete with the adoption of ironclad warships.

Modern Day Archaeological -The Penobscot Expedition Site

The Penobscot Expedition Site is a submerged historic archaeological area in the waters of the Penobscot River between Bangor and Brewer, Maine.

The area is the site of the abandonment and loss of many vessels in the disastrous 1779 Penobscot Expedition, an American Revolutionary War expedition in which the rebellious Americans lost an entire fleet of ships. The site was listed on the National Register of Historic Places in 1973; it has been of interest to salvagers and later archaeologists since the early 19th century.

Description and History

The Penobscot River flows into Penobscot Bay, a long bay that nearly bisects the state of Maine. At its head of navigation stand the cities of Bangor (on its western bank) and Brewer (on its eastern bank). The 1779 Penobscot Expedition was a military response by the state of Massachusetts (of which Maine was then part) to the seizure of Castine by British forces in June 1779.

Beset by poor leadership, the amphibious expedition was scattered by the arrival of a British fleet on the bay. All of the expedition's ships were captured, scuttled, burned, or abandoned. Nine armed vessels and as many as 16 transports are documented to have made it as far upriver as Bangor.

Soon after the expedition materials were salvaged by the Royal Navy from the abandoned and sunken ships.

Local residents salvaged readily accessible wrecks, and the state also authorized at least one formal salvage operation, the results of which are not known.

In 1809 Ebenezer Clifford recovered 30 cannons and several tons of cannonballs from the river. Later finds in the Penobscot River included cannons found on the river bottom in the Bangor-Brewer area in 1876 and in 1954-55. The obvious importance of the area, with a well-documented history, led to the area's listing on the National Register of Historic Places in 1973, primarily for the potential archaeological significance of materials located there, including military equipment, cargo, and other artifacts.

1779 Cannon retrieved from Penobscot River in 2000

Between 1994 and 1997 surveys conducted by the University of Maine located several wrecks in the Penobscot, tentatively identified as the *USS Warren* and the transport *Samuel.*

In 1998 Brent Phinney, the owner of a riverfront industrial business in Brewer, reported the presence of Revolutionary War-era finds near his property in Brewer, and opposite downtown Bangor.

These prompted archaeological teams from the United States Navy (which retains an interest in military shipwrecks) to conduct preliminary surveys in 1999, and more detailed fieldwork and excavation in 2000 and 2001.

These surveys determined that the Phinney Site on the Brewer side was of a shipwreck, and that the Shoreline Site on the Bangor side consisted of dispersed artifacts, including cannon and other military hardware. The ship was determined to be a two-masted brig or schooner, and has tentatively been identified as the privateer *Diligent*.

The privateer Diligent

References

Want of Proper Spirit and Energy - The Penobscot Expedition of 1779: Burbank, Major Dale W.

A People's Army: Massachusetts Soldiers and Society in the Seven Year's War. Anderson, Fred.

Crucible of War: The Seven Years' War and the Fate of Empire in British North America, 1754-1766. New York: Random House.

The Life and Surprising Adventures of John Nutting, Cambridge Loyalist and his Strange Connection with the Penobscot Expedition of 1779: Batchelder, Samuel F.

Documentary History of the State of Maine, vol 12. Portland, ME: Lefavor-Tower Co., 1908: Baxter, James P.,

Documentary History of the State of Maine, vol 16.-17 Portland, ME: Lefavor- Tower Co., 1910.

Soldiers in King Philip's War. Boston: Rockwell and Churchill Press, 1896: Bodge, George Madison.

The Penobscot Expedition: Commodore Saltonstall and the Massachusetts Conspiracy of 1779. Annapolis: Naval Institute, 2002: Buker, George E.

Maine at Louisburg in 1745. Augusta, ME: Burleigh and Flynt, 1910: Burrage, Henry.

The Siege of Penobscot. 1781; repr., New York: Arno Press, 1971: Calef, John .

The History of Philip's War, Commonly Called the Great Indian War, of 1675 and 1676: Church, Benjamin, Thomas Church, and Samuel Gardner Drake.

Naval Documents of the American Revolution, vol 2. Washington: U.S. Government Printing Office, 1966:   William Bell,

Ships and Seamen of the American Revolution. New York: Dover Publications, 1969: Coggins, Jack

Creating Portland: History and Place in Northern New England. Lebanon, NH: University of New Hampshire Press, 2005: Conforti, Joseph .

United States State-Level Population Estimates: Colonization to 1999. Fort Collins, Colorado: US Department of Agriculture, 2003: Coulson, David P. and Linda Joyce.

Changes in the Land. New York: Hill and Wang, 1983: Cronon, William.

Key to a Continent. Englewood Cliffs, NJ: Prentice-Hall, 1965: Downey, Fairfax. Louisbourg

The Border Wars of New England. New York: Charles Scribner's Sons, 1897.

The Taking of Louisburg, 1745. Boston: Lee and Shepard, 1890.

The Present State Of New-England With Respect To the Indian War. Boston: Samuel N. Dickison Printer, 1833.

Rise, and Fight Again: Perilous times along the Road to Independence. New York: Dodd & Mead

The History of Colonel Jonathn Mitchell's Cumberland County Regiment: Bagaduce Expedition: 1779. Portland, ME: Press of the Thurston, 1899: Goold, Nathan.

General Solomon Lovell and the Penobscot Expedition 1779. Weymouth, MA: Weymouth Historical Society, 1976.

Flintlock and Tomahawk; New England in King Philip's War. New York: Norton, 1966.

Revolution Downeast: The War for American Independence in Maine. Amherst: University of Massachusetts, 1993.

Correspondence of William Shirley, Governor of Massachusetts and Military Commander in America, 1731-1760, vol 2. New York: Macmillan,

Lovell, Solomon and Gilbert Nash. The Original Journal of General Solomon Lovell Kept during the Penobscot Expedition, 1779: with a Sketch of His Life by Gilbert Nash. Weymouth, MA: Weymouth Historical Society, 1881.

Maine Historical Society. Collections and Proceedings of the Maine Historical Society, vol 2, part 4. Portland, ME: Maine Historical Society, 1891.

Dispatch from General Francis McLean to General Henry Clinton about the Penobscot Expedition Forces,‖ 23 August 1779.

Dispatch from General Henry Clinton to General Francis McLean on the Penobscot fort. Maine Society of the Sons of the American Revolution. Maine at Valley Forge. Portland, ME: 1908.

Massachusetts. The Acts And Resolves, Public And Private, of the Province of the Massachusetts Bay: To Which Are Prefixed The Charters of the Province, vol 3. Boston: Albert J. Wright, 1878.and vol 21 1908

The Life of Lieutenant-General Sir John Moore. London: John Murray, 1834: Moore, James Carrick.

Military Affairs in North America, 1748-1765. New York: D. Appleton-Century Company, 1936: Pargellis, Stanley.

A Half Century of Conflict, vol I. and II Boston: Little, Brown and Company, 1892: Parkman Francis,

France and England in North America, A Series of Historical Narratives, Part

Fifth. Boston: Little, Brown, and Company, 1882. Peckham, Howard Henry.

The Colonial Wars: 1689-1762. Chicago, IL: University of Chicago, 1964.

Records of the Governor and Company of the Massachusetts Bay in New England, vol. 1. 1628-1641. Boston: William White Press, 1853: Shurtleff, Nathaniel Bradstreet

Records of the Governor and Company of the Massachusetts Bay in New England, vol. II. 1642-1649. Boston: William White Press, 1853.

"New Ireland: Men in Pursuit of a Forlorn Hope, 1779-1784,": Maine Historical Society Quarterly, 1979, Vol. 19 Robert W. Sloan,

The Lobster Coast. New York: Viking/Penguin: Woodard, Colin.

"The Search for Security: Maine after Penobscot," Maine Historical Society Quarterly James S. Leamon

Sir John Moore at Castine. Maine Historical Society.

A History and Description of New England. Boston, Massachusetts: Coolidge, Austin J. (1859).

The Envy of the American States: The Loyalist Dream for New Brunswick (1984): Ann Gorman Condon,

"Sir John Coape Sherbrooke". Dictionary of Canadian Biography Online. University of Toronto.

The Proposed Province of New Ireland. Collections of the Maine Historical Society 1904: Joseph Williamson.

John Calef Memorials and Petitions, 1766-1782

Records of the Governor and Company of the
Massachusetts Bay in New England, vol. 5. 1674-
1686. Boston: William White Press, 1854.

Castine Past and Present: The Ancient Settlement
of Pentagoet and the Modern Town. Boston:
Rockwell and Chuchill Press, 1896: Wheeler,
George Augustus.

History of Castine, Penobscot and Brooksville.
Bangor, ME: Burr and Robinson, 1875.

The Colonial Laws of Massachusetts: Reprinted
from the Edition of 1672, With the Supplements
Through 1686. Boston: Rockwell and Churchill
City Printers, 1890: Whitmore, William Henry

Wright, Robert K. Jr. Massachusetts Militia Roots:
A Bibliographic Study. Washington, DC: U.S.
Government Printing Office, 1989.

Zelner, Kyle F. A Rabble in Arms: Massachusetts
Towns and Militiamen during King Philip's War.
New York: New York University Press, 2009.

Barnes, Viola F. —The Rise of William Phips: The
New England Quarterly (July 1928): 290-296.

The Militia of Colonial Massachusetts. Military
Affairs 18, no 1: Radabaugh, Jack S.